Wenchuan, Sichuan Province, China, Earthquake of 2008

Lifeline Performance

PREPARED BY

Earthquake Investigation Committee of the
Technical Council on Lifeline Earthquake Engineering

EDITED BY

Alex K. Tang, P.E., P.Eng., C.Eng.

Technical Council on Lifeline Earthquake Engineering
Monograph No. 39

Published by the American Society of Civil Engineers

Library of Congress Cataloging-in-Publication Data

Wenchuan, Sichuan Province, China, earthquake of 2008 : lifeline performance / prepared by Earthquake Investigation Committee of the Technical Council on Lifeline Earthquake Engineering ; edited by Alex K. Tang, P.E., P.Eng. ; Technical Council on Lifeline Earthquake Engineering.
pages cm. — (Monograph ; no. 39)
Includes bibliographical references and index.
ISBN 978-0-7844-1333-3 (print : alk. paper)—ISBN 978-0-7844-7864-6 (ebook) 1. Buildings--Earthquake effects—China—Wenchuan Xian. 2. Lifeline earthquake engineering—China--Wenchuan Xian. 3. Earthquake damage—China—Wenchuan Xian. 4. Wenchuan Xian (Sichuan Sheng, China)—Buildings, structures, etc. I. Tang, Alex, editor. II. American Society of Civil Engineers. Earthquake Investigation Committee.
TH1095.W465 2014
693.8'52095138—dc23

2014023610

Published by American Society of Civil Engineers
1801 Alexander Bell Drive
Reston, Virginia, 20191-4382
www.asce.org/bookstore | ascelibrary.org

Errata: Errata, if any, can be found at http://dx.doi.org/10.1061/9780784413333.

ISBN 978-0-7844-1333-3 (paper)
ISBN 978-0-7844-7864-6 (PDF)
Manufactured in the United States of America.

Cover photo credits: top left: David Lee; top right: Conrad Felice; bottom left: John Eidinger; bottom right: Alex Tang

TCLEE Monographs

Since 1990, the Technical Council on Lifeline Earthquake Engineering (TCLEE) has published reports in the TCLEE Monograph Series. TCLEE Monographs further the Council's goal of advancing the state-of-the-art and -practice in lifeline engineering for earthquakes, hurricanes, and other extreme events.

More information about TCLEE is available at http://www.asce.org/tclee/.

A list of available TCLEE Monographs may be found at http://www.asce.org/bookstore. Digital versions may be found in the ASCE Library, http://ascelibrary.org/r/tclee.

Contents

Preface

The Earthquake Investigation Committee of the Technical Council of Lifeline Earthquake Engineering (TCLEE), American Society of Civil Engineers (ASCE), was established to initiate, organize, train for, coordinate, and evaluate the performance of lifelines following earthquakes. Members of the committee are employees of lifeline industries, consulting engineers, and academics from the United States and Canada. Committee members provide services on a voluntary basis. For some earthquake investigations, participants' companies do not require their employees to take vacation time for the investigation and may provide some support for expenses. ASCE also provides support to reimburse expenses. In addition to the time associated with the reconnaissance trip, the substantial effort by each individual to prepare a short report for the TCLEE Web page and the full report for the monograph series is all done on a voluntary basis. The cost of this effort is substantially more than the support provided by ASCE.

Individuals participating in the investigation need not be members of the committee or members of ASCE, but they are expected to follow the committee's earthquake investigation practices as described in the ASCE publication, TCLEE Monograph 11, *Guide to Post-Earthquake Investigation of Lifelines*. Members of the investigation team coordinate with other groups and may participate in groups organized by other organizations. They gather both good and poor performance data from domestic and foreign earthquakes to provide information for practitioners to improve the performance of the lifeline systems. Foreign earthquakes that have been investigated include the 1985 Chile, 1988 Soviet Armenia, 1990 Philippines, 1991 Costa Rica, 1992 Kocaeli (Turkey), 1994 Kobe (Japan), 1999 Kocaeli (Turkey), 1999 Chi-Chi (Taiwan), 2001 Gujarat (India), 2002 Atico (Peru), 2004 Zemmouri (Algeria), 2004 Sumatra, 2007 Kashiwazaki (Japan), and 2008 Pisco (Peru). Domestic earthquakes that have been investigated include 1989 Loma Prieta, 1992 Landers, 1994 Northridge, 2000 Napa, 2001 Nisqually, 2002 Denali, 2003 Paso Robles, and 2008 Alum Rock.

The Kobe earthquake report, Monograph 14, was the first foreign earthquake investigation report published by ASCE as a TCLEE monograph. The first domestic earthquake investigation report published by ASCE as a TCLEE monograph, Number 8, was for the Northridge earthquake. Prior to this time, TCLEE prepared a lifeline report that was published by the Earthquake Engineering Research Institute (EERI). The Earthquake Investigation Committee continues to cooperate with EERI to provide an abbreviated version of lifeline performance in Earthquake Spectra (EERI publication). TCLEE publishes brief preliminary reports on the ASCE/TCLEE Web page.

Alex Tang, PE, F. ASCE
June 2010

Contributors

Name	Affiliation	Email
John Eidinger	G&E Engineering Systems, Inc.	eidinger@earthlink.net
Curt Edwards	PSOMAS	cedwards@psomas.com
Conrad Felice	Washington State University	cfelice@wsu.edu cfelice@lachel.com
David Lee	EBMUD	dlee@ebmud.com
Diarmuid MacNeill	Dolmen Structural Engineers Inc	diarmuid@dolmen-engineers.net
Del A. Shannon	Black and Veatch	ShannonSA@bv.com
Jun Yang	University of Hong Kong	junyang@hku.hk
Mark Yashinsky	Caltrans	mark_yashinsky@dot.ca.gov
Phil Yen	FHWA	Wen-huei.Yen@fhwa.dot.gov
Alex Tang	L&T Consultant	alexktang@mac.com

ASCE/TCLEE Phase I Team (15 July – 19 July, 2008)

Conrad Felice David Lee
Jun Yang Alex Tang (team leader)

Mark Yashinsky a member of the Federal High Way Authority (FHWA) team visited highway bridges in July 2009.

ASCE/TCLEE Phase II Team (18 Oct – 23 Oct, 2008)

John Eidinger Curt Edwards
Del A. Shannon Alex Tang (team leader)

ASCE/TCLEE Lifeline Workshop Team invited by the Department of International Affairs, Sichuan Association for Science and Technology (25 May–29 May, 2009)

Curt Edwards Conrad Felice
Philip WH Yen Alex Tang (team leader)

ASCE/TCLEE Lifeline Workshop hosted by Sichaun Association for Science and Technology (20 August – 23 August 2012) meeting with the Zipinpu Development Company and the Operations staff and engineers

Curt Edwards John Eidinger
Diarmud MacNeill Alex Tang (team leader)

Acknowledgments

Most of the contributing authors of this report are members of the Earthquake Investigation Committee (EIC) of Technical Council on Lifeline Earthquake Engineering (TCLEE). Curt Edwards, John Eidinger, and Alex Tang were funded by ASCE. David Lee was funded by EBMUD. Conrad Felice was funded by Lachel Felice & Associates. Del Shannon was partially funded by ASCE and by Black & Veatch. Jun Yang was partially funded by ASCE and by University of Hong Kong. Mark Yashinsky was funded by EERI. The lifeline workshop team members of May 2009 were funded by ASCE (except Conrad Felice of WSU) and local expenses were funded by the Department of International Affairs, Sichuan Association of Science and Technology. ASCE support was received at the direction of Pat Natale, ASCE executive director, with support from John Durrant, John Segna, Tenzing Barshee, and Mike Sanio. The investigation team is very grateful to Conrad Felice who introduced Prof. Manchao He of China University of Mining & Technology, Beijing, as the primary contact in China, providing us with letter of invitations and obtaining site access approval from local government. Prof. He is now a member of the China Academy of Science (2013) the highest honor of research scientists in China.

In addition to all the individuals acknowledged in the chapters, many groups and individuals in the field also assisted the investigation team. In particular those individuals who set up the debriefing session organized by Prof. He prior to the field trip proved to be extremely helpful during the days in the field. Prof. He, who had visited the earthquake impacted areas prior to our trip, joined us throughout the whole investigation. He looked after all the fine details of field logistics for the team including accident insurance for the team members. He also provided the team with his photo collections to enrich the monograph. We were deeply indebted to his generous and dedicated support of the ASCE/TCLEE investigation.

We were thankful for Prof. Lizhong Yang of Southwest Jiaotong University, Sichuan, and his Ph.D. student Jian Wang, who provided the Phase I team with local logistics support, including the approval letter from the provincial government to pass through check points. Mr. Wang accompanied the team in the field as our interpreter.

We were also greatly indebted to Prof. Zhirong Mei, Board chairman and professor of the China Railway Southwest Research Institute, who graciously provided the Phase II team with field access as well as accompanying the team. He also provided us with the facility to host a workshop on lifeline earthquake in Chengdu.

We were also very grateful that Director Junwei Cui of the Department of International Affairs, Sichuan Association for Science & Technology, organized a meeting with the Sichuan Electric Power to review and collect the electric power system performance data. Later in May 2009, he also helped the team to access restricted areas after the workshop he organized in Chengdu. Without his support our trip would be only half as successful.

Local officials should be commended for the time they took from their busy restoration/relief efforts to provide us with the data and information contained in this report.

We appreciated the support of Jinfeng Zhou, Ph.D.; Maurice Strong; and Ray Zhou, all from Beijing. Their effort made the field investigation run smoothly.

Zifa Wang, Ph.D., of the Institute of Engineering Mechanics, China Earthquake Administration (IEM, CEA) provided us the earthquake records and his findings. Prof. Baitao Sun of the Institute of Engineering Mechanics, China Earthquake Administration (IEM, CEA) also provided us with his valuable information of damage in the area. Joseph Sun, Ph.D., of Pacific Gas and Electric provided us with interpretation of documents. J.P. Bardet and Tong Qiu of Geo-Engineering Earthquake Reconnaissance (GEER) provided us with geotechnical data.

As is not uncommon in post-earthquake reconnaissance, incomplete information in the weeks and months after the event can lead to omissions and misunderstandings. We apologize if the findings in this report are incomplete, and the reader is cautioned that it may take months to years of post-earthquake evaluations before a comprehensive understanding of lifeline impacts is available. ASCE/TCLEE will continue to update the lifeline performance information when more information is collected.

The findings and photos presented in this report reflect the collected input from many people and sources. The ASCE TCLEE team of Alex Tang, David Lee, Jun Yang, and Conrad Felice, collected site data from July 15 to July 19, 2008; Alex Tang, John Eidinger, Curtis Edwards, and Del Shannon collected data from October 18 to October 23, 2008; Alex Tang, Curt Edwards, Conrad Felice, and Phil Yen collected data from May 27 to May 28, 2008.

Alex Tang, P. Eng., C. Eng., P. Log., F. ASCE, Team Leader.

All of the team members express great support for Alex Tang for his tireless leadership, patience, and dedication, without which this report would not have happened.

John Eidinger PE, M. ASCE, on behalf of the team

This monument is located at the city entrance to Dujiangyan, about 30 km from fault rupture. It depicts a flying horse. Its head and tail are raised high in a proudly untrammeled gallop, its fleeting touch with the earth suggested in the one hoof borne on a flying swallow. It is patterned after bronze horses discovered in 1969 at the tomb of General Li Ghangli, dating back to the second century BC. We want to reflect that this particular earthquake disaster is only one of many during the long history of China and that the resiliency of Chinese society will allow the affected communities to be rebuilt to honor the dead and move on to a prosperous and proud future.

Prof He arranged a meeting with professors from Qinhua University and from China University of Mining and Technology the first day in Beijing. Front row from left to right - Zhang Shizhong (张世忠), Xie Qiao(谢峤), Manchao He (何满朝), Alex Tang, Yu Weiping(于卫平). Back row from left to right - Yang Xiaojie(杨晓杰), Kong Weili(孔维利), Zhao Lian(赵廉).

Prof. He organized a debriefing session in Beijing for the team prior to the investigation trip. From left to right, Alex Tang, Prof Manchao He, Conrad Felice, David Lee, and Jun Yang.

From left to right, Roy Zhou, secretary of Jinfeng Zhou (周晋峰), Honorable Maurice Strong, and Alex Tang at the debriefing session.

Alex Tang presented a plate of appreciation from ASCE to Prof He (何满朝) for his generous support to the investigation team.

Phase I team group photo in front of the Jinqilin Hotel in Chengdu before heading out to the earthquake impacted area on July 17 2008. From left to right, Conrad Felice, Jun Yang, David Lee, Alex Tang, He Manchao, Jian Wang.

A group photo of Prof He's students (standing) with the investigation team after a tour of the rock mechanics laboratory in Beijing.

Post investigation lecture set up by Prof He and attended by his students.

David Lee (on the left) joined Prof He's post graduate students at lunch after the post investigation lecture.

Phase II investigation team group photo with Sichuan host, Mei Zhirong Board Chairman of China Railway Southwest Research Institute (second from right). Left to right. John Eidinger, Curt Edwards, Alex Tang, Manchao He, Mei Zhirong (梅志荣), and Del Shannon in front of the rock that fell from the mountain on the left. It is too big to move, and it is close to the epicenter, therefore it becomes the monument. Words on the rock – May 12 epicenter, Yingxiu.

The attendees at the Chengdu workshop; the banner read, "Welcome the Experts from ASCE."

The Chengdu workshop – ASCE/TCLEE

Phase II investigation team members with the leaders of the China Railway Southwest Research Institute after the morning session of the workshop. From left to right Luo Chaoting (罗朝廷), Wan Xiaoyan （万晓燕）, Del Shannon,Yang Lizhong (扬立中)，Curt Edwards, Li Lin （李林）, He Manchao(何满朝), Alex Tang (邓国权),John Eidinger, Mei Zhirong （梅志荣), Yan Jinxiu （严金秀), Ye Guorong（叶国荣），Su Jian （粟健） (Courtesy of China Railway Southwest Research Institute).

Director Cui (Junwei) (崔俊伟) (far left) organized the information session with Sichuan Electric Power Company for the investigation team.

1 INTRODUCTION

EXECUTIVE SUMMARY

The epicenter of Wenchuan M 7.9 earthquake of May 12, 2008 was in Yingxiu at N31.0° latitude, E103.4° longitude, about 100 km northwest of Chengdu. The earthquake occurred at 2:28 p.m. local time.

The earthquake caused widespread and extensive damage to critical infrastructure lifelines. Among the most affected were high-voltage substations and transmission lines, highway bridges, highway tunnels, roads, railroad bridges and tunnels, dams, communication centers, and water systems. The total number of fatalities ranged from 60,000 to 90,000, including about 6,000 students from primary to high schools.

The primary contributing effects to the damage were the following:

1. Seismic design requirements were not established in this area until the late 1990s. Buildings and equipment installed prior to that time had no seismic requirements.
2. The seismic design rules established in the late 1990s were grossly deficient for this type of earthquake. The design level peak ground acceleration was only 0.1g, and the response spectra, ductility assumptions for non-ductile styles of construction, and importance factors were all inconsistent with prudent design for M 7+ earthquakes.
3. The actual ground motions were commonly PGA $\geq$ 0.3g within 10 km either side of the primary fault ruptures, with the duration of strong ground shaking (PGA > 0.1g) commonly 40 seconds or longer. There were some instrumented recordings with PGA exceeding 0.8 g and ground velocities of 50 in. per second.
4. Many landslides, rock falls, and debris flows contributed to the damage of roads, tunnel portals, dams, and the general building stock.
5. The quality of construction per se did not appear to be a contributing factor to the widespread destruction; given the seismic design implemented, the performance of the buildings and equipment due to strong ground shaking was just about as expected—terrible.

Power Systems

Electric power generators and the transmission grid to the provinces are owned by the central government in China. Each province has its own power company to handle local distribution. Both the state grid and the Sichuan power company sustained extensive damage, particularly in the mountainous regions.

Damage to Sichuan Electric exceeded 400 equipment component failures at more than 100 substations (110 kV—500 kV), including damage to more than 100 power transformers. Almost all of the damage was from inertial overloads. The 110 kV to 230 kV transmission lines in mountainous areas within 10 km either side of ruptured faults suffered many tower failures due to rockfalls and deep-seated landslides.

Communications

Overall the communication system performed poorly. There was damage to equipment in the central offices and cell sites. The distribution system in the unfriendly mountainous terrain sustained extensive damage from collapsed bridges, landslides, and rock falls. The outage in the remote areas lasted as long as 4 weeks.

Prolonged electric power outage was a major factor of the communications system's poor performance.

Remote areas were inaccessible due to road failures, collapsed bridges, landslides, and rock falls. This caused longer recovery duration.

More than 600 telecommunication equipment buildings sustained damage—ranging from total collapse to cracks on walls. Approximately 6,800 km of optical fiber cables were damaged. There were a few cases of cellular towers collapsed. One of the service providers reported a direct loss of ¥3.48 B (RMB) (approximately US$530 million).

Water and Wastewater Systems

The city of Dujiangyan had a water supply problem. The other smaller towns and villages depend on spring water from the mountain.

Highways and Bridges

The highways and bridges are one of the lifelines that sustained the most damage. This hindered emergency response and rescue. Due to landslides, rockfalls, and bridge failures, access to towns and villages in the mountainous areas became very difficult and time consuming. When heavy equipment delivered by the military opened roads, the rescue teams and emergency response resources could reach towns and villages with building collapses and lifelines services disruptions.

Approximately 30,000 miles of roads and railways, 3,000 bridges, 100 tunnels, and many miles of retaining structures were damaged by the earthquake. Losses to the transportation sector exceeded US$10 billion.

Railways and Tunnels

Many passenger and cargo trains were stranded on railway tracks linking Chengdu to the rest of the country. Many railway tracks were damaged by permanent ground deformation, landslides, and rockfalls. A significant number of tunnels (including highway tunnels) were damaged. The most common damage was from landslides or rockfalls blocking tunnel portals. About 90 percent of the total length of one highway tunnel leading to Yingxiu was damaged by collapses, heaved road surface, and the lining falling off.

Airports

The international airport in Chengdu, Shuanliu (双 流) Airport, escaped damage. The main terminal was shut down for a short period of less than 20 minutes due to cellular phone service interruption. Incoming air traffic continued to land as normal.

The airport was then closed to schedule flights a few days later to accommodate relief supply flights including military personnel. It reopened for all scheduled flights, including international flights, by the end of May.

Dams

There are more than 7,000 dams in Sichuan Province. About 70 percent of these were built in the 1950s and 1960s. On May 25, 2008, according to Chinese Ministry of Water Resources, there were 2,380 dams operating under emergency conditions due to earthquake-induced damage; 69 of these were reported to be in danger of failing. Another report suggests that 18 dams nearly collapsed, and 135 dams were severely damaged.

As of October 20, 2008, the team knew of no gross failures of any dams (release of water), but we did observe damage. Although there was no severe damage to dams within the impacted area, many reservoirs lowered water levels to reduce potential hazard down stream.

In addition, landslides and debris flow blocking rivers in the mountainous areas created many earthquake-formed dams.

Schools

Unfortunately, student fatalities reached 6,000. Many factors contributed to the collapsed or partially collapsed schools. It is easy for a trained structural engineer working in the seismic design of buildings to observe the damage of reinforced concrete buildings and report that there were insufficient reinforcement bars, poor detailing of connections, lateral load carrying capacity, and such. For unreinforced masonry (URM) buildings, the conclusion is simply that URM should not be used in a high seismic zone. All these conclusions are valid; however, the answer is much more basic than that: Firstly, the zoning of this region is grossly under estimated. Secondly, the building code does not emphasize the importance of schools being designed to a high lateral resistance. Thirdly, some of the construction and material quality is questionable. In addition, the inspection processes are not clearly specified.

When the buildings are not designed to resist high seismic load, coupled with poor construction and material quality and the absence of an effective inspection processes, it is not surprising that so many schools collapsed. These issues exist in many countries. Even North America has issues with older school buildings in high seismic zones.

Emergency Response

Many emergency response lessons will be learned from this mega disaster facing a government in peacetime; however, one lesson above all is mitigation.

Due to extensive road, highway, and bridge damage, accessing the hardest hit areas in this mountainous terrain became the biggest challenge. Before heavy construction equipment was delivered, the first military personnel sent in to provide emergency response had to walk hours to reach victims.

The People Liberation Army (PLA) of China should be highly commented for their dedication, responsiveness, and hard work to ease the pains of the victims in this disaster.

The estimated number of people affected by this earthquake is about 4.6 million. Hundreds of thousands of tents and temporary housing were set up for the victims. Thousands of tons of supplies—food, water, daily necessities, and medical supplies—were delivered by trucks and helicopters.

Monograph Organization

This report discusses the following lifeline systems: electric power, communication, water and wastewater, highways and bridges, airports, railways, schools and building stock, and emergency response. Most chapters provide an overview of the system's performance, followed by sections describing the system, its damage, emergency response, and recovery. Each chapter finishes with conclusions and recommendations.

Lessons Learned

The damage observed to lifelines in the Wenchuan earthquake is likely to be repeated in other locations in future Chinese earthquakes, for all locations assigned Intensity VII (as was all of Wenchuan area) or even VIII (Chinese design scale).

If China adopts modern seismic design guidelines for its infrastructure, such as those commonly adopted in the high seismic zones of California, it is likely that more than 90 to 95 percent of all infrastructure damage will be avoided in future earthquakes in China.

China should upgrade its seismic design criteria to at least Intensity IX (Chinese scale) or PGA = 0.30g for all portions of the country that are exposed to M 6+ events more than once every 100 to 200 years. Even higher requirements (PGA = 0.40g) should be established within 10 km on both sides of active faults. Zoning requirements should prohibit the construction of buildings for human occupancy beneath steep slopes subject to earthquake or intense-rain triggered landslides and rockfalls. The "importance factor" in the building code should be 1.0 or greater, reducing the risk of under design.

China should implement a seismic retrofit program for its entire infrastructure. The retrofit program should include high-priority, low-cost fixes (for example, anchorage of all equipment), and selectively upgrade buildings (such as schools and hospitals), and facilities (such as lifeline equipment buildings and control centers) for the effects of inertial shaking. All new critical infrastructures must be designed for the simultaneous effects of strong ground shaking, landslide, liquefaction, and surface fault offset.

Mitigation is the key in reducing future losses of human lives and property because it is certain that earthquakes will happen there again.

Investigation Team

The Post Earthquake Investigation Committee of the Technical Council on Lifeline Earthquake Engineering (TCLEE), a technical council of the American Society of Civil Engineers (ASCE) organized two reconnaissance teams, the Phase I and Phase II teams, to perform reconnaissance of the lifeline in the earthquake area. The Phase I investigation team started the reconnaissance effort on July 6, 2008, and completed the initial efforts on July 10, 2008. The Phase II team operated from October 18 to October 22, 2008. Using two teams reduced the number of individuals in the disaster area, a request of the local emergency response official. The investigation teams consisted of the following ASCE members:

Phase I Team

Alex Tang (team leader of Phase I, Phase II, and May 2009 Lifeline Workshop), L&T Consultant, alexktang@mac.com

David Lee, EBMUD, dlee@ebmud.com

Jun Yang, University of Hong Kong, junyang@hku.hk

Conrad Felice, Washington State University, cfelice@wsu.edu (also on Lifeline Workshop May 2009 team)

Phase II Team

John Eidinger, G&E Engineering Systems, Inc., eidinger@earthlink.net

Curtis Edwards, PSOMAS, cedwards@psomas.com (also on Lifeline Workshop May 2009 Team)

Del A. Shannon, Black and Veatch, ShannonSA@bv.com

Mark Yashinsky, Caltrans, mark_yashinsky@dot.ca.gov, team member of the FHWA group, agreed to contribute to the transportations chapter

Lifeline Workshop Team of May 2009

Alex Tang, Conrad Felice, Philip W. H. Yen (Wen-huei.Yen@fhwa.dot.gov), and Curt Edwards.

Lifeline Workshop Team of August 2012

Alex Tang, John Eidinger and Curt Edwards.
The team also visited the Zipingpu Dam, which is very close to the epicenter of the 2008 earthquake.

Chapter Authors

The chapter authors are as follows:

The Front Matters, Alex Tang and John Eidinger

Chapter 1 (Introduction), John Eidinger, and Alex Tang

Chapter 2 (Seismology and Geotechnical), John Eidinger, Conrad Felice, Jun Yang, and Alex Tang

Chapter 3 (Transportations), Mark Yashinsky, John Eidinger, and Alex Tang

Chapter 4 (Electric Power), John Eidinger, Eric Fujisaki, and Alex Tang

Chapters 5 (Water and Wastewater), David Lee and John Eidinger

Chapter 6 (Communications), Alex Tang

Chapter 7 (Airports), John Eidinger and Alex Tang

Chapter 8 (Dams), Del Shannon, John Eidinger, and Alex Tang

Chapter 9 (Schools and General Building Stock), John Eidinger, Alex Tang, Conrad Felice, and Jun Yang

Chapter 10 (Emergency Response), Curt Edwards, Alex Tang, and Jun Yang

Eric Fujisaki of Pacific Gas & Electric Company visited some substation in the region, and agreed to contribute his findings to Chapter 4 of this monograph.

Even though every attempt was made to verify the accuracy of the information in this report, due to the nature of post-earthquake investigations this report may still contain inaccuracies and/or incomplete data. The authors apologize in advance to all those people who will point out new, updated, or corrected findings and observations.

Maps, Place Names, and Locations

The towns and cities visited by the ASCE team included Chengdu (gateway city); Dujiangyan City, Yingxiu, Hongkou, Nanba Town, Hejiaba Town, Jiangyou Town, Mianzhu Town, Shifang Town, and Beichuan Town. Town is used to refer to populations between 30,000 and 200,000 people. City refers to populations larger than 300,000 people. Village refers to small developments of a few hundred to several thousand people.

Several place names in China use various English spellings depending on the source and map. Most available maps of the areas are shown with most towns listed only in Chinese. To help interpretation, we note below the various spellings used for the same place in this report, along with latitude and longitude coordinates. In most places, this report uses the first noted spelling indicated below, but alternate spellings used in other reports and source materials are shown. Next to the place names, we also reference observed MMI intensities, using MMI IX to represent the highest rate of damage due to ground shaking and MMI X, XI exclusively to describe damage due to moderate to extreme effects of landslide. MMI values for X, XI, and XII do not correspond to increasing PGA values.

Anxian, An Xian

Beichuan (MMI IX, XII)

Chengdu (MMI IV - VI)

Deyang

Dkang

Dujiangyen (MMI V - VII)

Ertaishan (substation in Yingxiu) (MMI IX)

Hangkou (MMI IX, X)

Jiangyou

Jujiaya

Nanba

Pengzhou City

Shenxigou Village (near Hangkou) N 31° 05.895' , E 103° 37.959'

Shifang

Xiaoba

Yingxiu, Yingxou (MMI IX)

Yuanmenba

Abbreviations

MMI	Modified Mercalli Intensity
M	Magnitude (moment)
PGA	Peak Ground Acceleration (g)
TCLEE	Technical Council on Lifeline Earthquake Engineering
WTP	Water Treatment Plant

2 SEISMOLOGY AND GEOTECHNICAL

2.1 Seismic Setting

The Wenchuan, China M 7.9 earthquake is one of the most damaging in the last several hundred years in a country that has documented many devastating earthquakes during the past 700 years. Table 2.1 lists estimated fatalities for earthquakes with magnitude 8 or larger and other similar events.

Table 2.1. Fatalities in Past Major Earthquakes in China

Year	Location	M	Deaths
1303	Shanxi	8	200,000
1411	Tibet	8	N/A*
1556	Shaanxi	8.5	830,000
1654	Gansu	8	31,000
1668	Shandong	8.5	50,000
1679	Hebei	8	45,000
1739	Ningxia	8	50,000
1812	Xinjiang	8	58
1833	Tibet	8	5
1833	Yunnan	8	6,707
1879	Gansu	8	30,000
1902	Xinjiang	8.3	5,650
1920	Taiwan	8	5
1920	Ningxia	8.5	235,000
1927	Gansu	8	40,000
1931	Xinjiang	8	300
1933	Sichuan	7.3	8,300
1950	Tibet	8.6	3,300
1951	Tibet	8	N/A*
1976	Tangshan	7.8	655,000[1]
2008	Wenchaun (Sichuan)	7.9	89,000

** not available*

Table 2.2 lists additional significant earthquakes in the region since 1917. The locations of these events, along with the epicenter for the May 12, 2008 earthquake, are shown in Figure 2.1.

[1] Fatality estimate based on *National Geographic, Earthquake Risk, a Global View* (2006). The fatality estimate by Chinese authorities was 242,000.

Table 2.2. Regional Earthquakes with M > 7

Year	Month	Day	Time	Latitude	Longitude	Depth	Magnitude
1917	07	30	2354	29.000	104.000	0	7.3
1923	03	24	1240	30.553	101.258	25	7.2
1933	08	25	0750	31.810	103.541	25	7.3
1947	03	17	0819	33.000	99.500	0	7.5
1948	05	25	0711	29.500	100.500	0	7.2
1950	08	15	1409	28.500	96.500	0	8.6
1955	04	14	0129	29.981	101.613	10	7.5
1967	08	30	0422	31.631	100.232	8.1	7.0
1973	02	06	1037	31.361	100.504	6.6	7.4

Fig. 2.1. Regional tectonics (Courtesy IEM)

Figure 2.2 shows the regional tectonic environment. The epicenter of the May 12, 2008, Wenchuan earthquake is shown by the large dark oval. The earthquake originated on the Longmenshan thrust zone (Fig. 2.2). It was manifested by breakage of two (likely three) faults in near succession (Fig. 2.3), each striking northeasterly; the total fault rupture occurred over 300 km. The earthquake is the result of convergence of crust from the high Tibetan Plateau to the northwest against the strong and stable crust block underlying the Sichuan basin to the southeast. On a continental level (Fig. 2.1), the northward convergence of the Indian plate against the Eurasian plate of about 4 cm per year is broadly accommodated by the uplift of the Asian highlands and by the motion of crustal material to the east, away from the uplifted Tibetan plateau (Fig. 2.1).

Figure 2.3 shows the local tectonic province. The epicenters of the two main events in the May 12, 2008 earthquake are shown by the red dots. The depth of the earthquake was 12 to 19 km.

An estimate of the population exposed to various intensity levels is provided in Table 2.3. This table was developed by taking the approximate intensity map in Figure 2.4 and overlaying the assumed population density of Sichuan province, using population data available from Chinese sources. Field observations of damage and population densities during the team's 6 days in the field suggest that a more accurate estimate of population exposed to various levels of ground shaking might be closer to that in Table 2.4.

Figure 2.7 shows a Google Earth Satellite map of the region. The fault rupture propagated to the northeast. As can be seen, Chengdu is in a broad plain, with common elevation about 1,500 ft. above sea level. The fault rupture zones were near the base of the mountains about 80 km northwest of Chengdu. Figure 2.8 shows a close up of the region with fault rupture ("A" is the epicenter). The bulk of the area with intensity IX or higher can be characterized as mountainous with many valleys cut by rivers.

Figure 2.3 shows a cross sectional view of the region, which cuts through the three faults with surface rupture looking towards the northeast.

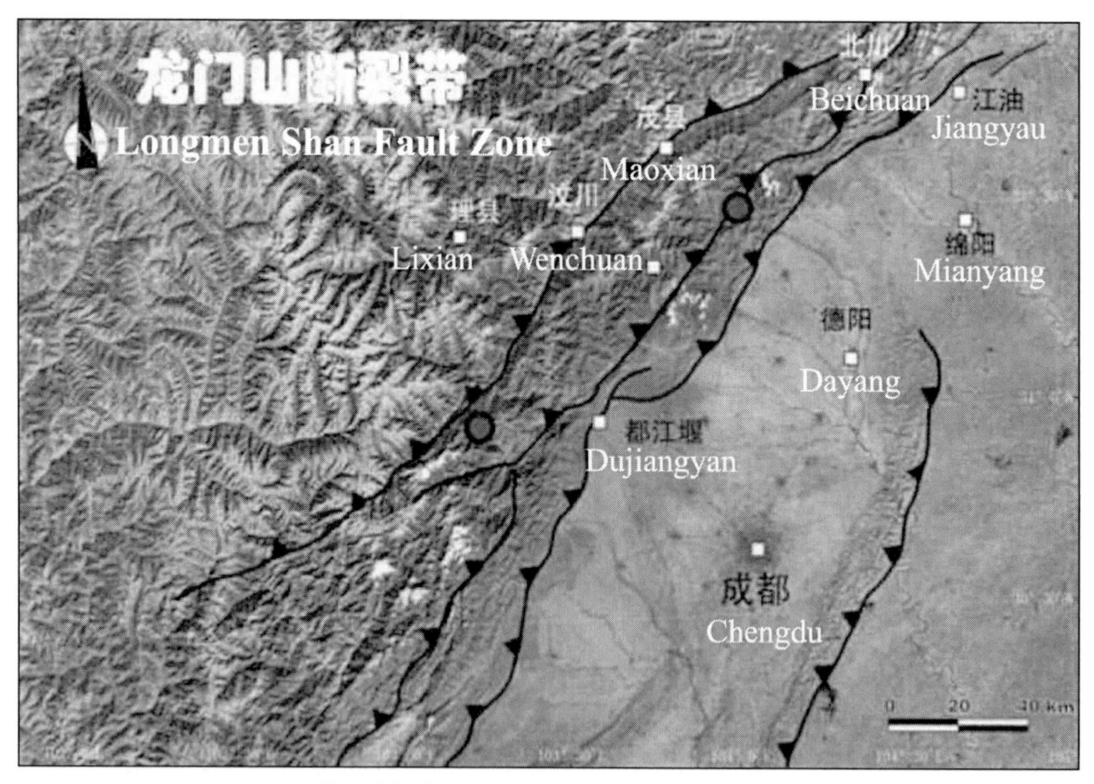

Fig. 2.2. Local tectonics (Courtesy IEM)

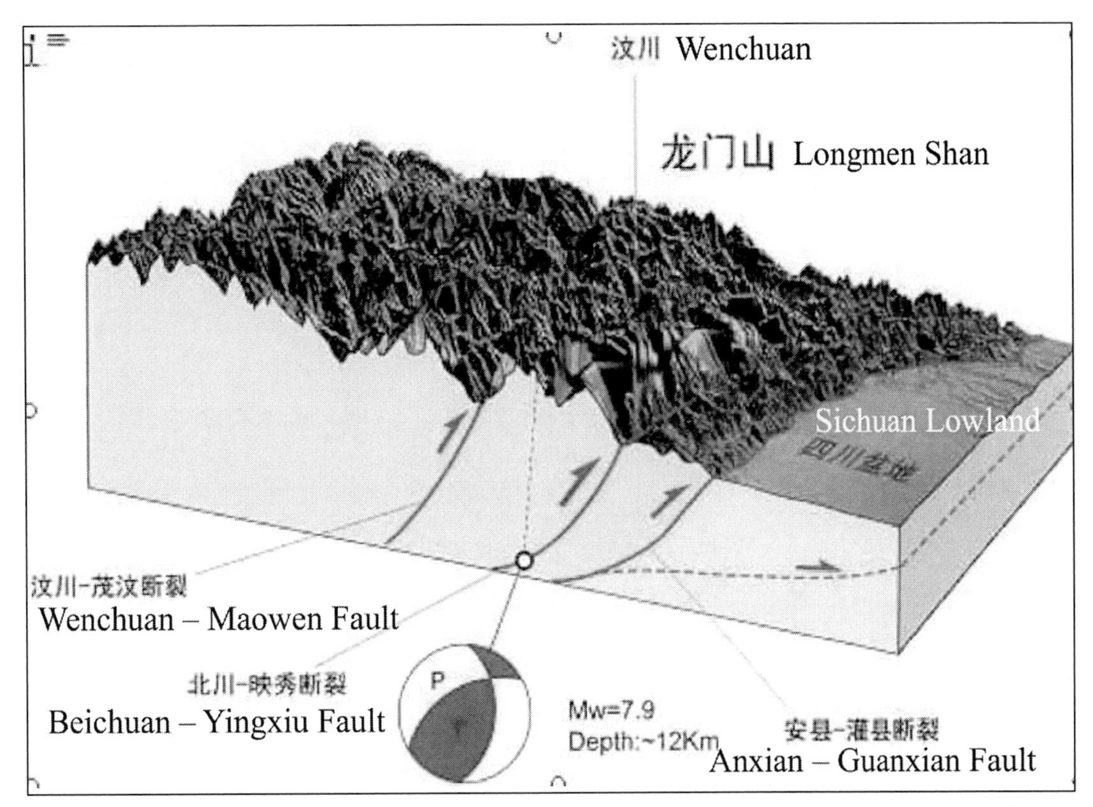

Fig. 2.3. Cross sectional view of the faults in the earthquake area (Courtesy IEM)

Figure 2.4 shows a shake-map with estimate of earthquake intensity. This map reflects the projection of the ruptured fault plane and is based on intensity-attenuation relationships; the intensity does not correspond to field-observed damage.

The bulk of the population in the mountainous valleys lives at elevations between 2,500 and 3,500 ft. above sea level. Within the first 30 km or so of the mountain range, the peaks are commonly at elevations of 3,000 to 6,000 ft. The slopes are 30 to 45 degrees, which is relatively steep. The subsurface materials observed (often at landslide cuts) were limestone and conglomerates. Rainfall in the area is common, and vegetation covers all mountain slopes except where denuded by the earthquake-triggered landslides. There were more than 12,000 rock falls/deep seated slides and more than 500 debris flows.

Figure 2.5 shows a map of the seismicity in the region since 1990. The epicenter of the May 12, 2008, event is shown by the large star close to center of the figure.

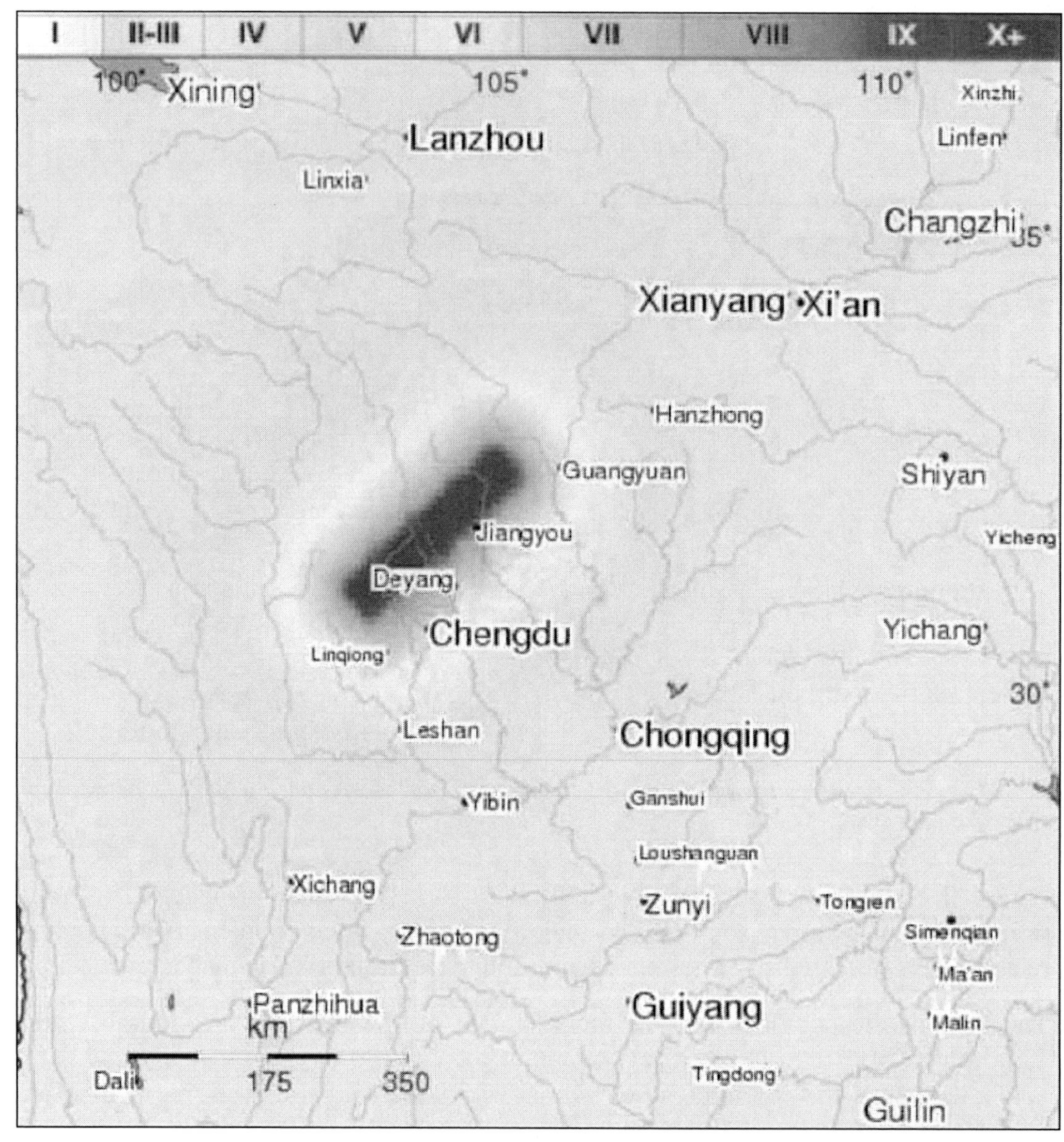

Fig. 2.4. Preliminary intensity map used for population exposure (Courtesy IEM)

Table 2.3. Population Exposed

Intensity	Population Exposed
X+	603,000
IX	692,000
VIII	4,301,000
VII	12,396,000
VI	15,484,000
V	89,480,000
IV	> 192,000,000

Table 2.4. Population Exposed (Field Estimated)

PGA	Population Exposed
≥ 0.50	250,000
0.30 – 0.50	250,000
0.15 – 0.30	3,000,000
0.05 – 0.15	14,000,000

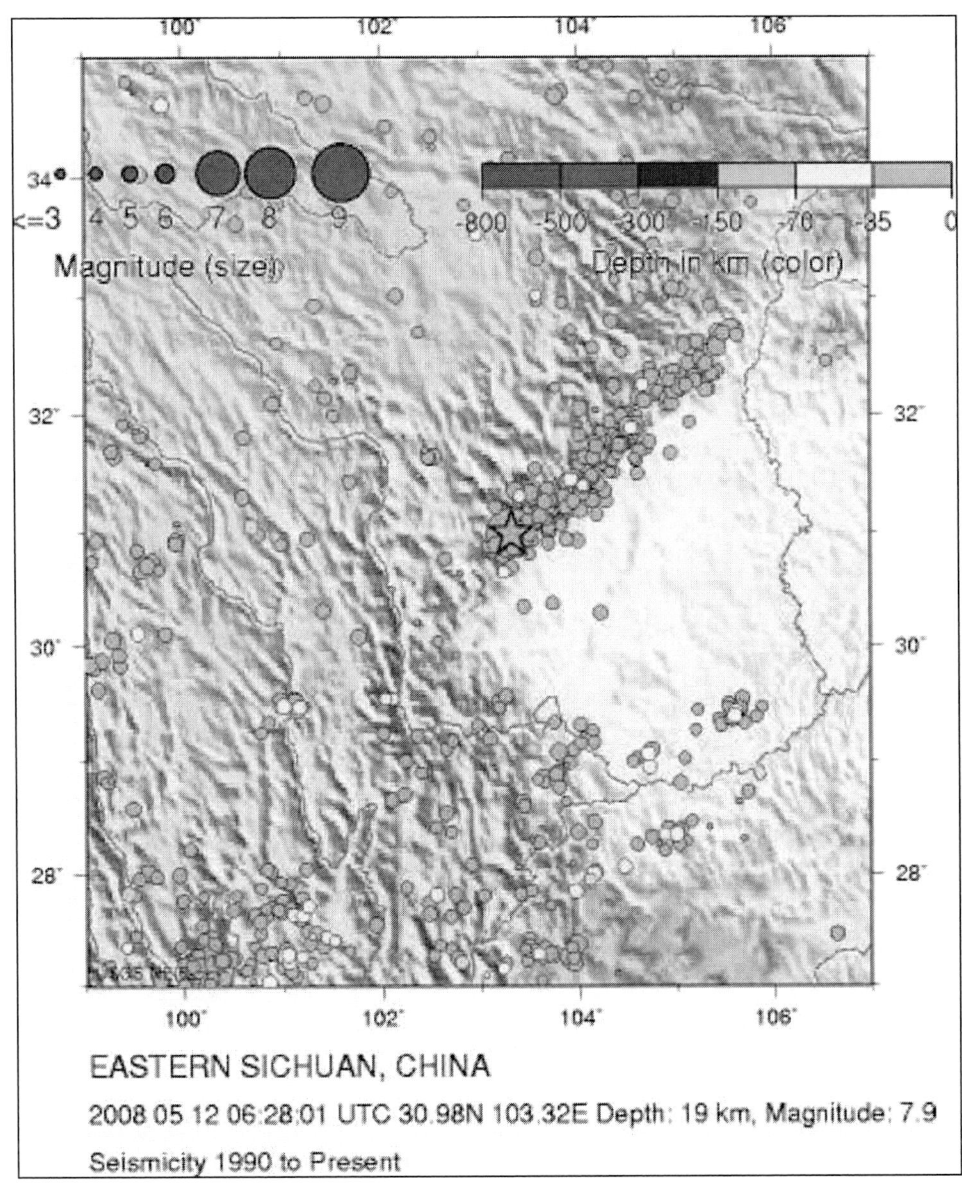

Fig. 2.5. Seismicity, 1990 to 2008 (Courtesy USGS)

Between the foothills of the mountains and Chengdu, the broad plain is populated by several moderate-size cities (100,000 to 500,000 population) intermixed by lightly industrialized areas and farming. This area experienced ground shaking commonly from 0.15 g to 0.25 g. Almost all damage in this area is from inertial shaking. The team observed some ground compaction issues, as evidenced by slightly settled sidewalks in urbanized areas. There may also have been liquefaction in some areas, as suggested by the high water pipe repair rates in Mianzhu City (PGA ~0.2 g); however, there was no surface faulting or landslides in this area. We observed evidence of several collapsed multi-story residential buildings in Dujiangyan (population over 250,000, PGA likely 0.20 g to 0.35 g city-wide). Immediately northwest of Dujiangyan are the famous water diversion works of Lu Bing, which experienced ground motions of 0.25 g to perhaps 0.35 g. Here, manmade riverbanks with retaining walls commonly experienced 2 to 3 in. of lateral and downhill movements (as observed for about 1,500 ft. of the 4,000 ft. of embankment that we walked along).

The mountainous zones visited by the ASCE team did not appear to have been glaciated within the last 100,000 years or so. There are infrequent snowfalls at the valley floors, and snow avalanches were not reported to occur in the areas visited. Had the local temperatures been somewhat cooler, resulting in yearly heavy snow accumulations with resultant avalanches, the building stock might have been located outside the most common steep slope areas. This would have reduced the number of lives lost due to earthquake-triggered landslides. Of relevance to parts of the world where large populations live in steep mountainous terrains with heavy snowfall is the potential for an earthquake-triggered landslide or avalanche, such as the ski-area canyons in Utah and California where M6.5 to M7 earthquakes are thought to have recurrence intervals on the order of several hundred to a few thousand years.

Figure 2.6 shows the surface traces where faulting reached the ground. The epicenter of the first rupture (Wenchuan-Maoxian fault) is in the southwest corner of the map; aftershocks (M 4.5 to M 6.4) are shown by white dots. The fault names reflect the names of local towns.

Based on our field observations, coupled with time history recordings from local strong motion instruments, we interpret the faulting sequence as follows. The Wenchuan-Maoxian fault ruptured first, followed by the Beichuan-Yingxiu fault and, finally, the Pengguan fault. This third event might have been largely sympathetic fault movement and might not have been associated with independent causation of major ground accelerations.

Figure 2.7 shows a three-dimensional view of the topographic relief, with the large yellow arrow showing the sense of rupture from Yingxiu in the southwest, towards the northeast.

Fig. 2.6. Faults that ruptured on May 12, 2008; "O" marks the epicenter (Courtesy of IEM)

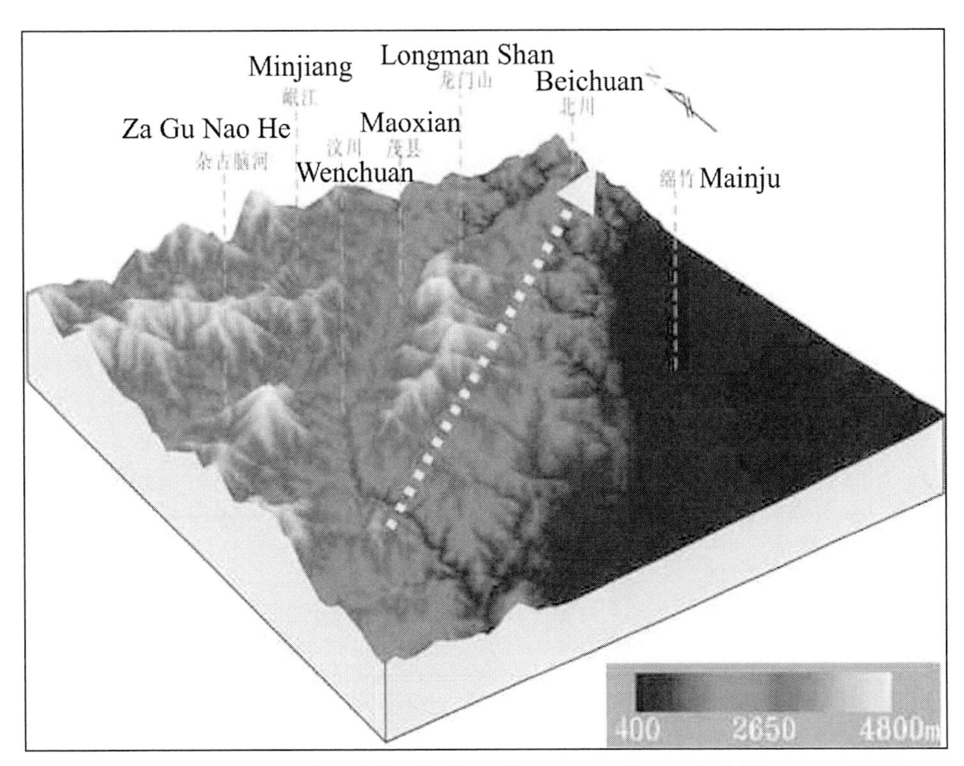

Fig. 2.7. Sense of strike of the fault and topographic relief (Courtesy IEM)

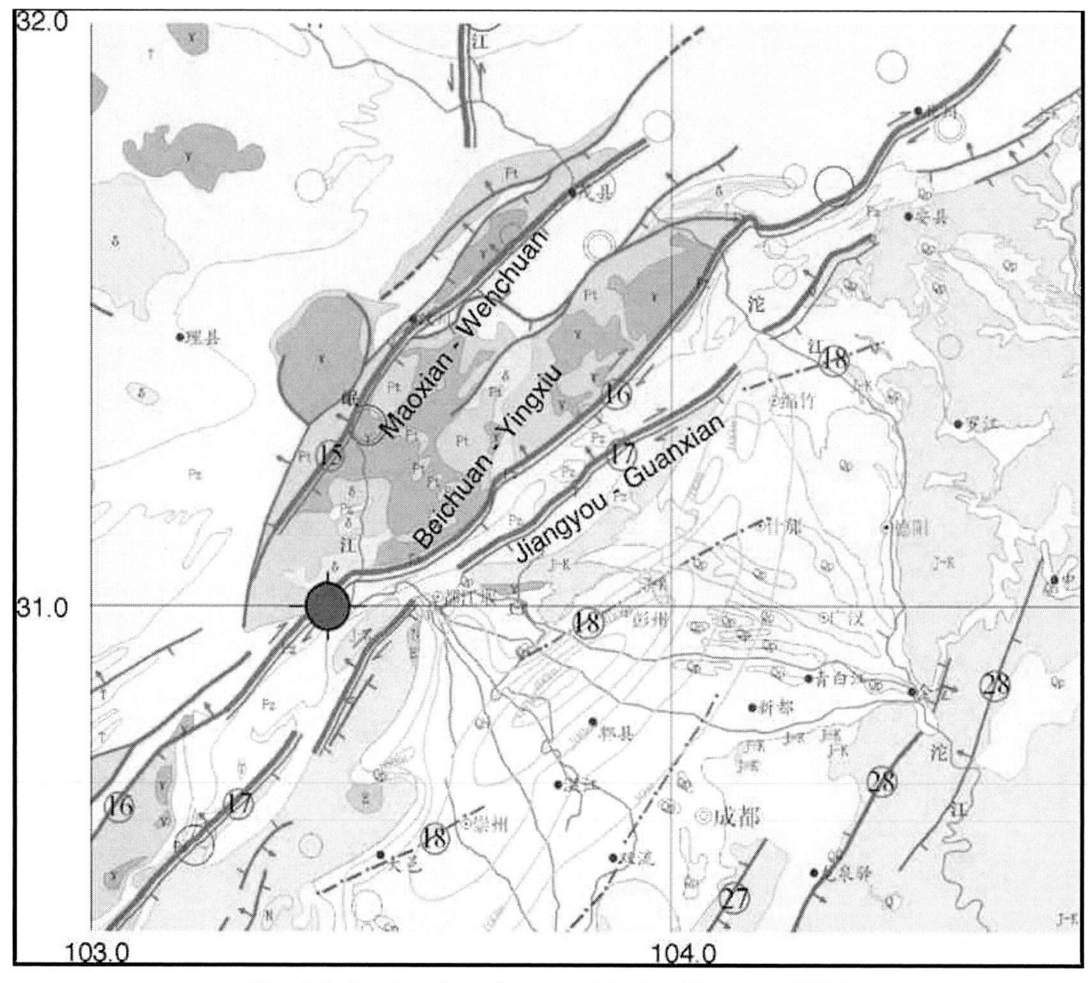

Fig. 2.8. Regional geology and faults (Courtesy IEM)

2.2 Seismic Hazard Zonation

Large earthquakes in the region had been recorded in 1610 and 1900, with more than 100 moderate to large earthquakes recorded since 1900 (see Table 2.2). This includes the following: Qinghai earthquake of 1900, Beichuan earthquake of 1913, Lixian-Maoxian M 7.3 earthquake of 1933, Maoxian earthquake of 1940, Kangding earthquake of 1941, Kangding earthquake of 1949, Kangding-Wenchuan earthquake of 1952, Beichuan earthquake of 1952, Dayi earthquake of 1970, and Mianzhu earthquake of 1999.

Burchfiel et al. (2008) provide research on the large-scale tectonic issues in the Longmenshan fault zone.

Figure 2.8 shows the greater Sichuan Provide area, highlighting mapped faults, surface geology on a regional level, and the epicenter of the May 12, 2008, event (large solid circle on the lower left).

The Chinese authorities have rated this area as a relatively low seismic hazard zone, and the design code for engineered structures suggested a design level of PGA =

0.10g (MMI VII) (see Fig. 2.10). Given the evidence of past earthquakes from Table 2.2 and Figure 2.9, the seismic zoning map of Figure 2.10 requires updating to reflect the true hazard of the region. The strong ground shaking of the May 12, 2008, earthquake, and the observed damages and loss of lives should be the ultimate evidence to update the seismic zoning map (Fig. 2.10).

Figure 2.9 shows locations of historical records. The dots show the May 12, 2008, epicenter (large dark dot) and M6+ aftershocks (small dark dots). This lighter dots show historical earthquakes (larger dots for M 7+, smaller dots for M 6-6.9).

Informal discussions with various local engineering design agencies in Chengdu suggested that the de facto seismic design requirement was zero before the earthquake. As evidenced by the damage to various engineered structures (such as highway bridges, tunnel portals, communications buildings, and high-voltage substations), essentially no seismic design was incorporated into many lifelines. Apparently there were no landslide, surface faulting, or liquefaction zonation maps for the area available prior to the earthquake. The surface faulting that occurred went through many occupied buildings. Landslides were common and affected a great number of occupied structures.

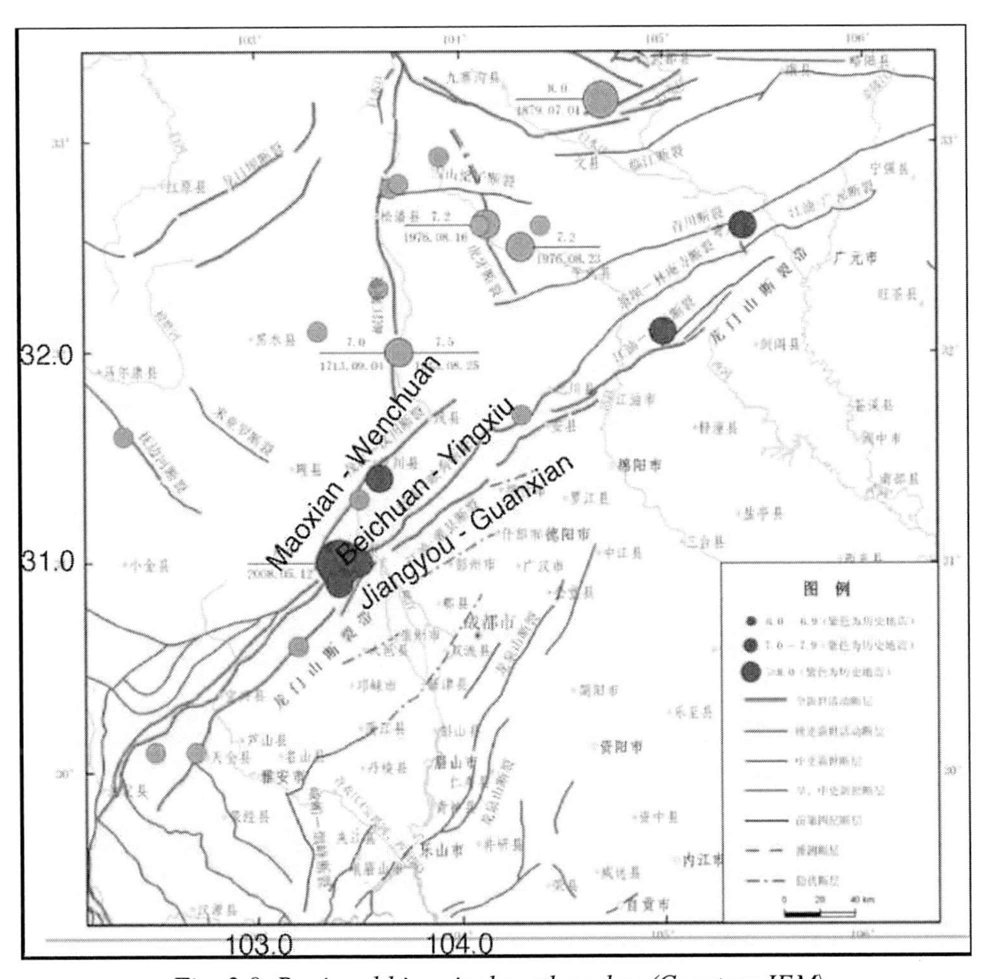

Fig. 2.9. Regional historical earthquakes (Courtesy IEM)

2.3 Ground Motions

The China Earthquake Administration (CEA) has installed some 2,000 strong motion instruments throughout China. As of mid-October 2008, the CEA had not released the digitized records to the community. Figure 2.11 shows the location of strong motion instruments in China; the earthquake zone is highlighted in an oval.

There were 19 recording stations within 100 km of the epicenter and 12 recording stations within 20 km of the ruptured fault. There were 120 records with PGA > 0.1 g.

Table 2.5 lists peak ground acceleration uncorrected motions for 20 instruments.

Table 2.6 provides the approximate locations of the instruments that recorded the strongest ground motions. The data in Table 2.6 were developed from available maps; the locations are likely not as accurate beyond one decimal point.

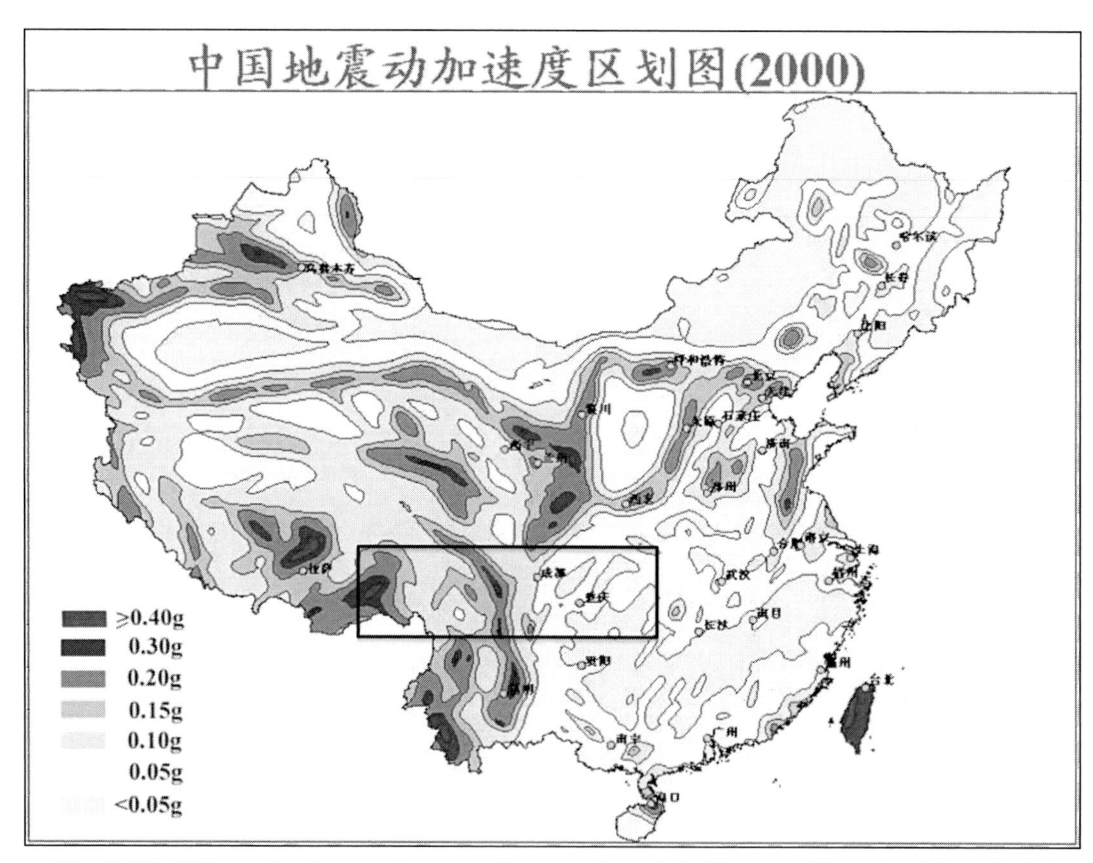

Fig. 2.10. China seismic zonation map (pre-earthquake) (Source: China Building Code)

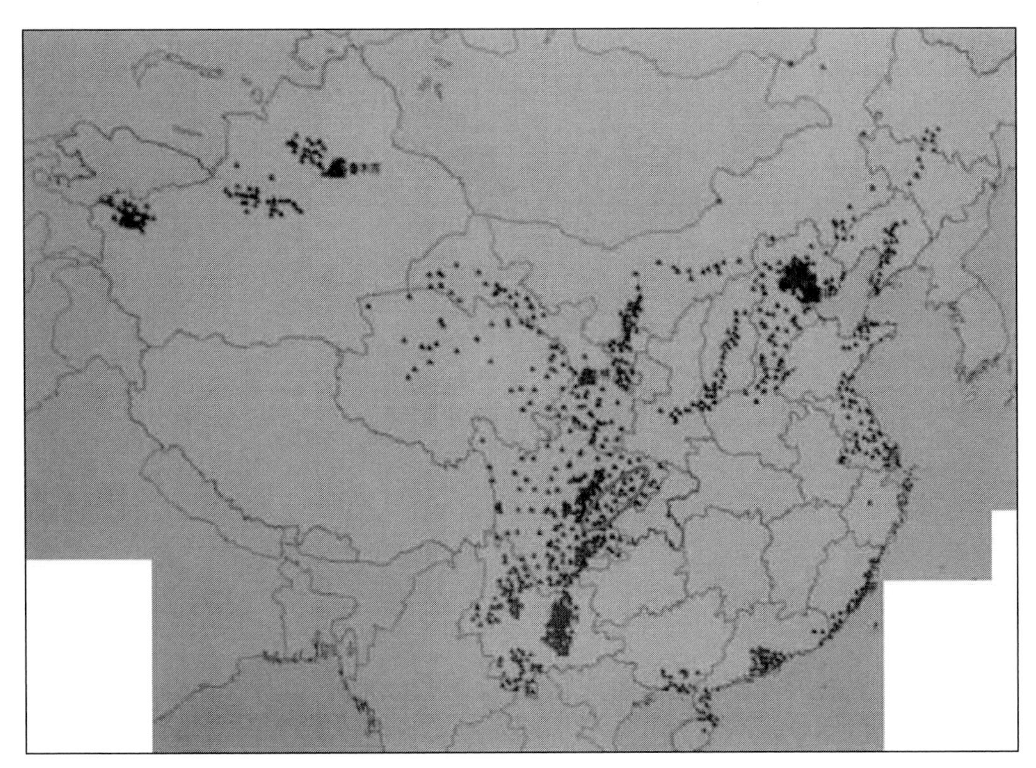

Fig. 2.11. China strong motion instruments (Courtesy CEA0)

Figure 2.12 shows the east-west direction recorded motion at station 51MZQ. This instrument is located near Beichuan, which is 0.74 km from surface faulting. At this site, the duration of very strong shaking (exceeding PGA = 0.20g) about 15 seconds, and duration exceeding PGA = 0.10g about 40 seconds. Peak motions were 828 gal and 133 cm/sec[2]. The duration of very strong ground shaking (PGA > 0.2g) was about 15 seconds, and moderately strong ground shaking (PGA > 0.1g) lasted about 55 seconds. The very large velocity pulse early in the record reflects, in part, that this recording station was at the end of the rupture opposite of the epicenter and is oriented approximately in the fault normal direction.

Figures 2.13, 2.14, and 2.15 show overlays of PGA contours on a regional scale; recording stations are indicated by black triangles. Figures 2.16, 2.17, and 2.18 show the same information but at a larger scale and with surface topography and major place names shown in the background. Contour intervals are 40 gal.

The instrument (051WCW) nearest the epicenter (22 km) recorded horizontal PGA = 0.97g. This instrument was 1.09 km from surface rupture.

[2] The waveform in Fig. 2.12 is very similar to the wave form that was used to develop the peak values in Table 2.5, so the minor differences (824 vs 828 gal) between peaks should be interpreted as data processing issues.

Table 2.5. Recorded PGA Values (981 gal = 1g)

Instrument Number	Station / Location	PGA EW gal	PGA NS gal	PGA Vert gal
051HSL	Shuangliusuo Stn in Heishui	107.64	142.561	108.97
051GYZ	Zengjia Stn in Guangyuan	424.480	410.480	183.338
051JYD	Jiangyou Stn	511.330	458.680	198.278
051JYH	Hanzeng Stn in Jiangyou	519.491	350.135	444.331
051JYC	Chonghua Stn in Jiangyou	297.187	278.961	180.489
051JZG	Guoyuan Stn in Jiuzhia	169.741	241.450	109.258
051LXM	Muka Stn in Li County	320.938	283.840	357.810
051LXS	Chaba Stn in Li County	221.260	261.755	211.088
051LXT	Taoping Stn in Li County	339.733	342.381	379.577
051MXT	Maoxian Earthquake Office	306.571	302.163	266.640
051MXD	Diexi Stn in Maoxian	246.493	206.210	143.910
051MXN	Nanxin Stn in Maoxian	421.281	349.239	352.480
051PWM	Muzuo Stn in Pingwu	273.744	287.370	177.370
051SFB	Bejiao Stn in Shifang	556.169	531.592	633.090
051WCW	Wolong Stn in Wenchuan	652.851	957.700	948.103
062WUD	Wudu Stn	184.874	163.990	108.610
051MZQ	Qingping Stn in Mianzhu	824.121	802.710	622.910
051DYB	Maima Stn in Deyang	126.290	136.326	88.990
051HYQ	Qingxi Stn in Hanyuan	142.330	125.190	54.415
051YAM	Mingsham Stn	171.160	175.370	46.693

Source: Institute of Engineering Mechanics Web site, www.smsd-iem.net.cn

Table 2.6. Station Locations

Instrument Number	Station / Location	Latitude (N)	Longitude (E)
051HSL	Shuangliusuo Stn in Heishui	33.45	105.00
051GYZ	Zengjia Stn in Guangyuan	32.65	106.20
051JYD	Jiangyou Stn		
051JYH	Hanzeng Stn in Jiangyou	31.75	104.62
051JYC	Chonghua Stn in Jiangyou	31.90	105.20
051JZG	Guoyuan Stn in Jiuzhia		
051LXM	Muka Stn in Li County		
051LXS	Chaba Stn in Li County		
051LXT	Taoping Stn in Li County	31.50	103.50
051MXT	Maoxian Earthquake Office	31.65	103.83
051MXD	Diexi Stn in Maoxian	32.10	105.80
051MXN	Nanxin Stn in Maoxian		
051PWM	Muzuo Stn in Pingwu		
051SFB	Bejiao Stn in Shifang	31.20	104.00
051WCW	Wolong Stn in Wenchuan	31.00	103.20
062WUD	Wudu Stn		
051MZQ	Qingping Stn in Mianzhu	31.52	104.09
051DYB	Maima Stn in Deyang		
051HYQ	Qingxi Stn in Hanyuan		
051YAM	Mingsham Stn		

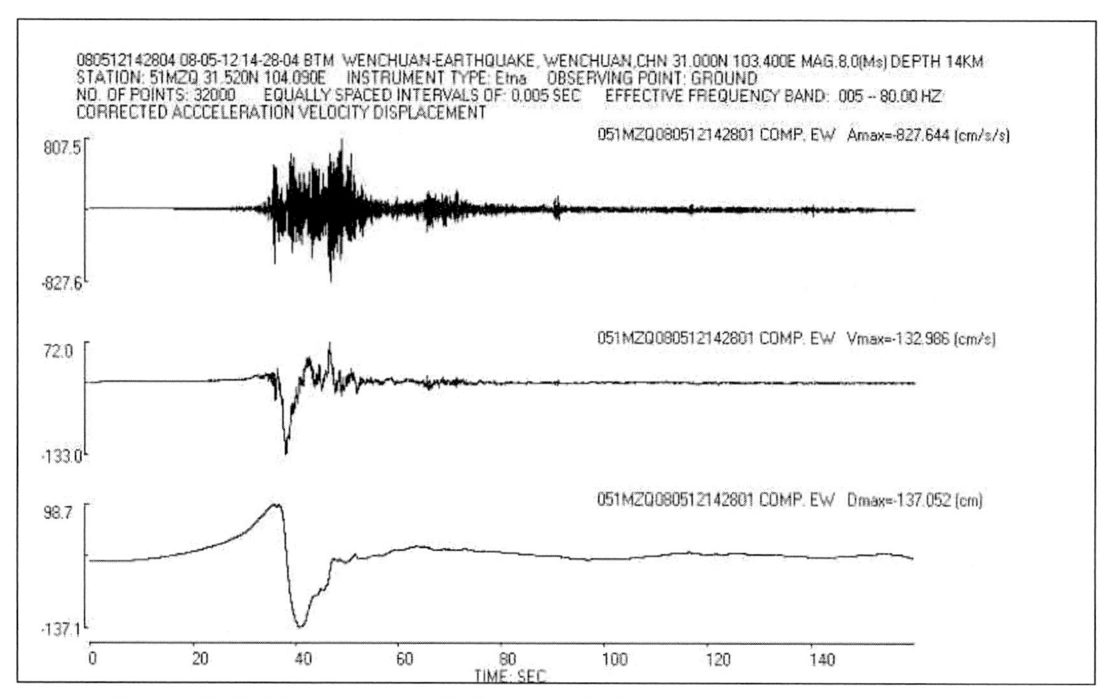

Fig. 2.12. PGA contours, E-W direction (gal, 981 gal = 1 g) (Courtesy IEM)

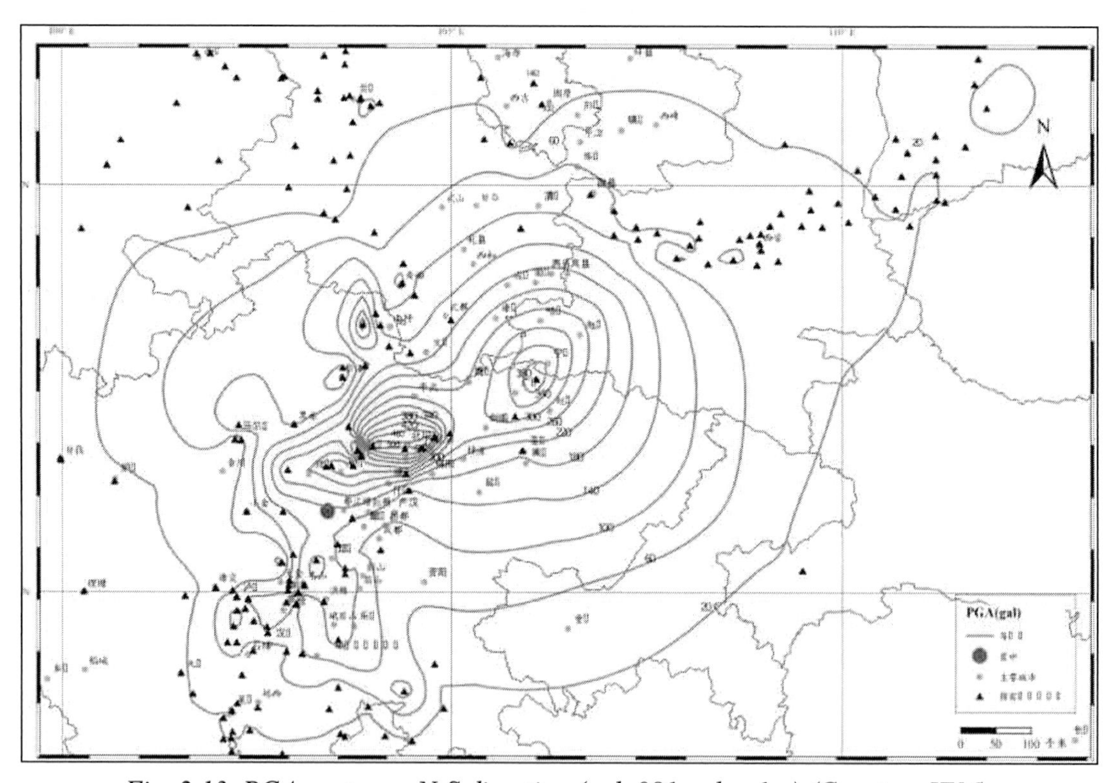

Fig. 2.13. PGA contours, N-S direction (gal, 981 gal = 1 g) (Courtesy IEM)

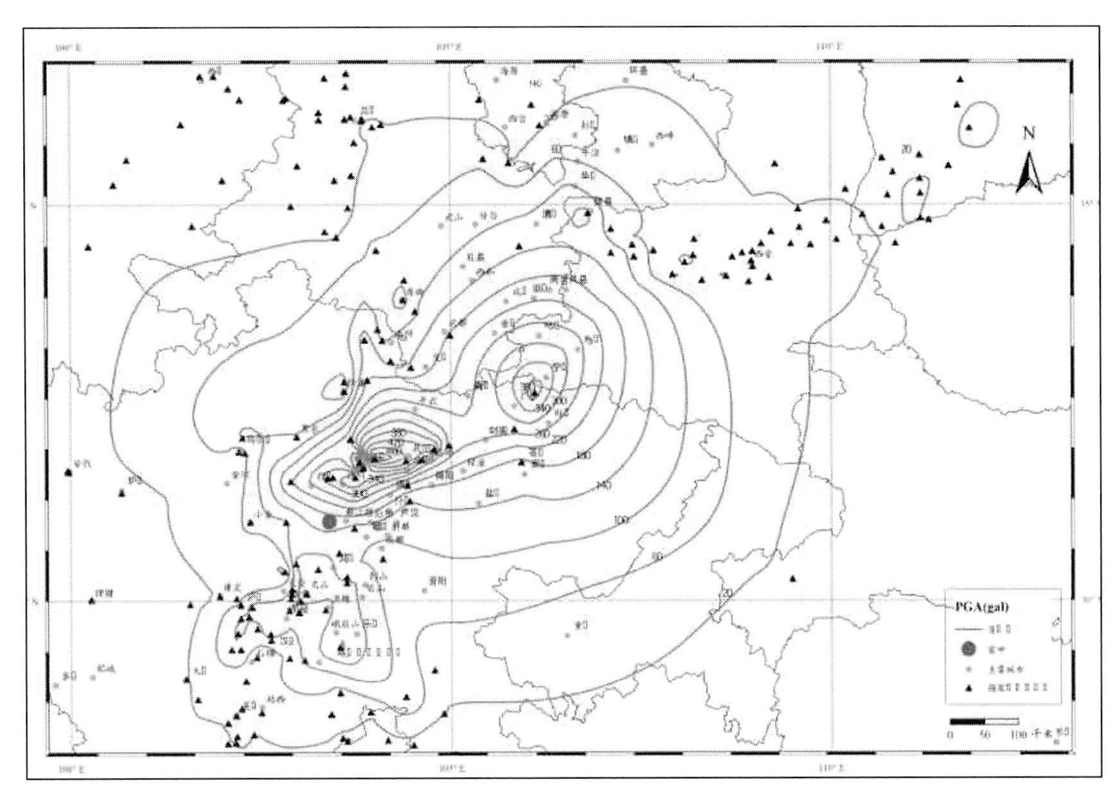

Fig. 2.14. PGA contours, E-W direction (gal, 981 gal = 1 g) (Courtesy IEM)

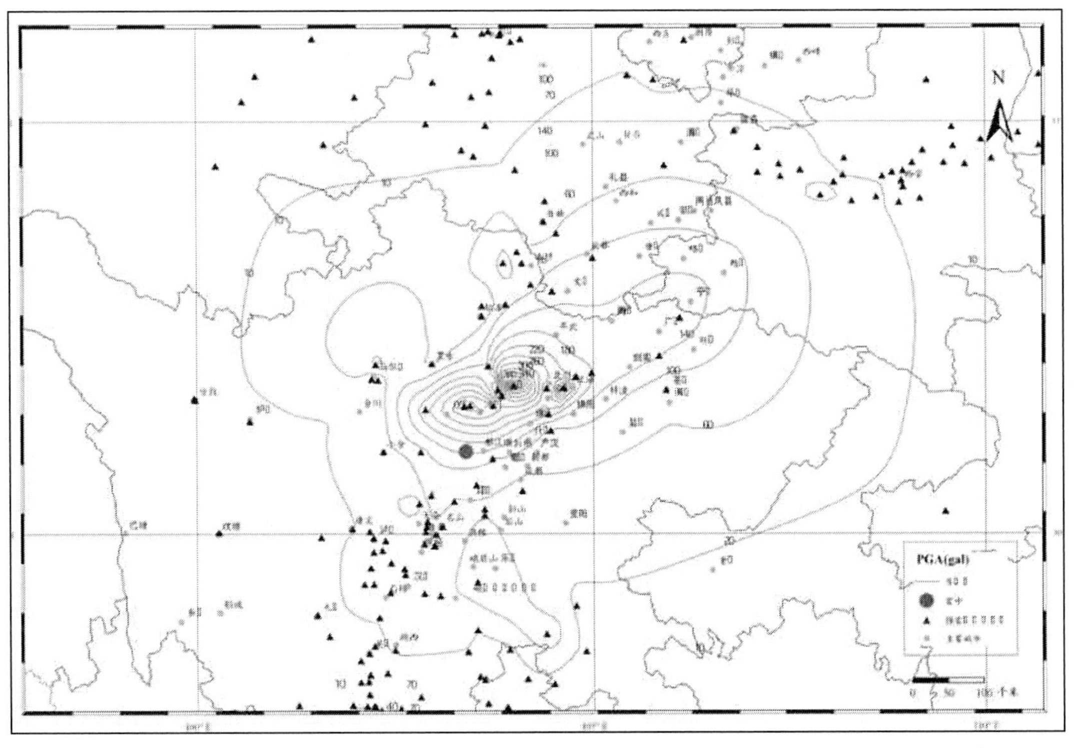

Fig. 2.15. PGA contours, vertical direction (gal, 981 gal = 1 g) Courtesy IEM)

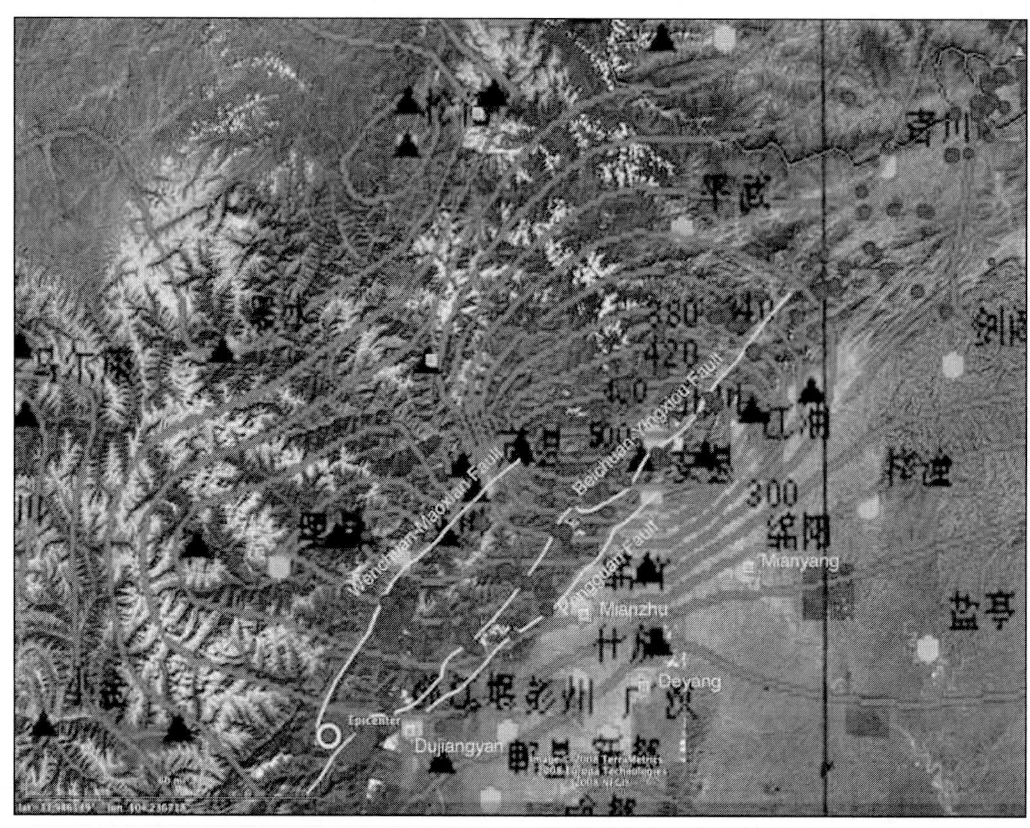

Fig. 2.16. PGA contours, N-S direction (gal, 981 gal = 1 g) (Courtesy IEM)

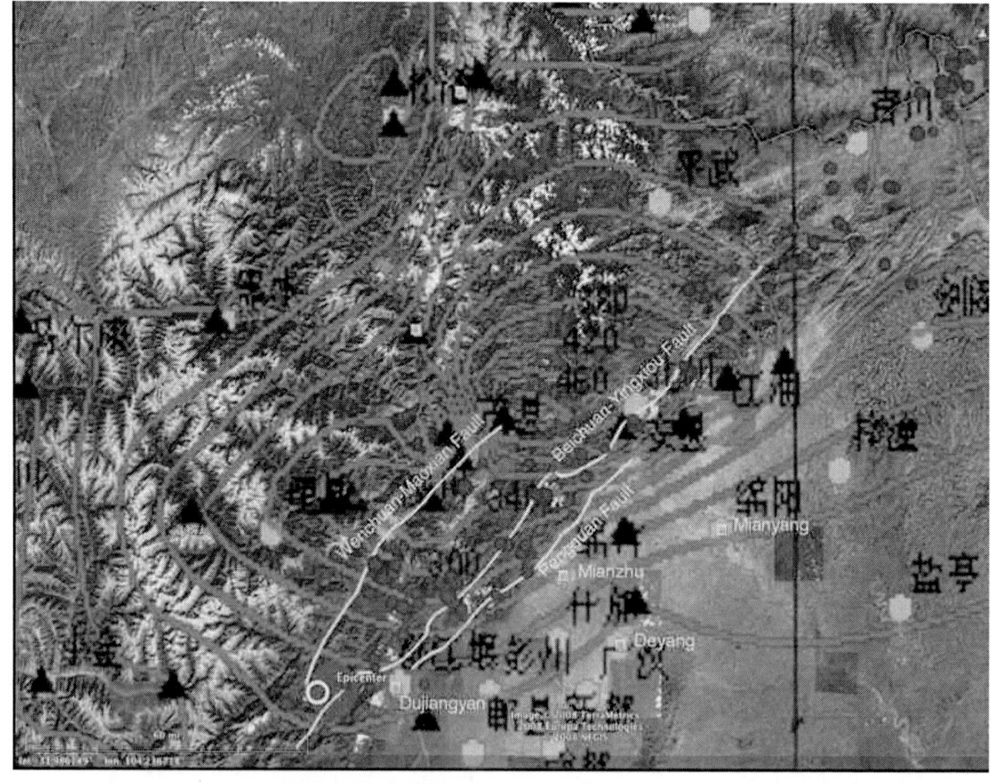

Fig. 2.17. PGA contours, E-W direction (gal, 981 gal = 1 g) (Courtesy IEM)

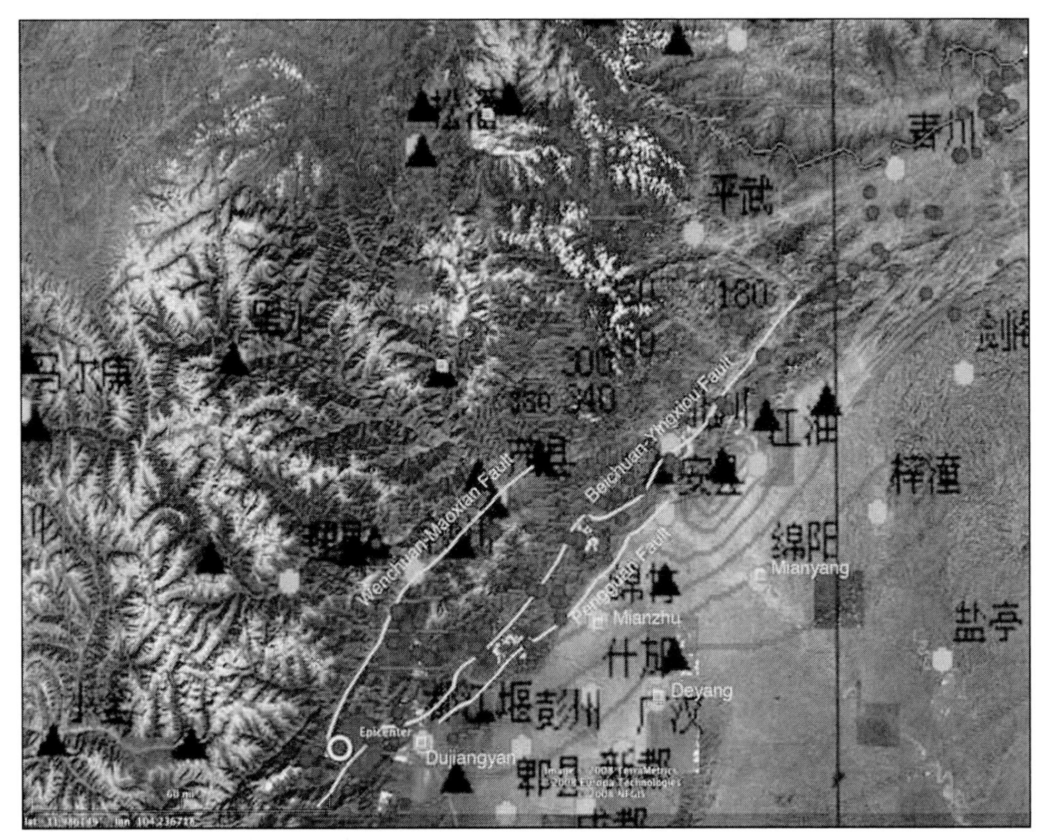

Fig. 2.18. PGA Contours, vertical direction (gal, 981 gal = 1 g) (Courtesy IEM)

Figure 2.19 is a section of the map shown in Figure 2.13, adding in the locations of the major cities in the Sichuan plain. The method of drawing the contours in the very near field of fault offset appears to be based only on using the raw data from Table 2.5 without consideration of the faulting zone. Based on field observations, we expect that PGA was much larger than 140 gal in the areas about 10 km northwest of Dujiangyan, but this is not reflected in the map.

Figure 2.20 shows interpreted MMI contours based on damage as observed by the CEA. The epicenter is shown as a ringed dot to the lower left of the darkest shaded zones. The darkest shaded zones reflect MMI XI (major landslide zones), the next darkest shaded area is MMI X, and the successive lighter shaded areas range from MMI IX to MMI VI. Most of Metropolitan Chengdu is shown as MMI VI, but the observations of the ASCE team suggest that a more detailed map for Chengdu would show most of the metropolitan area as MMI IV or V, with pockets of VI.

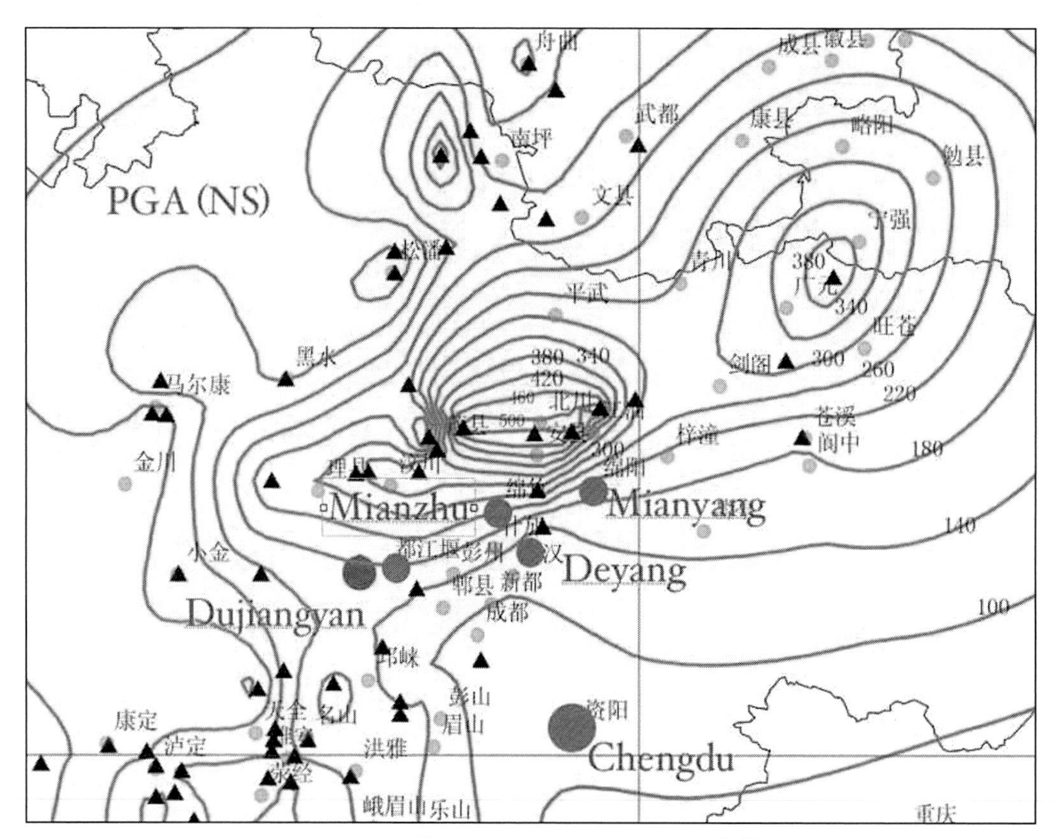

Fig. 2.19. PGA contours, NS direction, with major place names (gal) (Courtesy IEM)

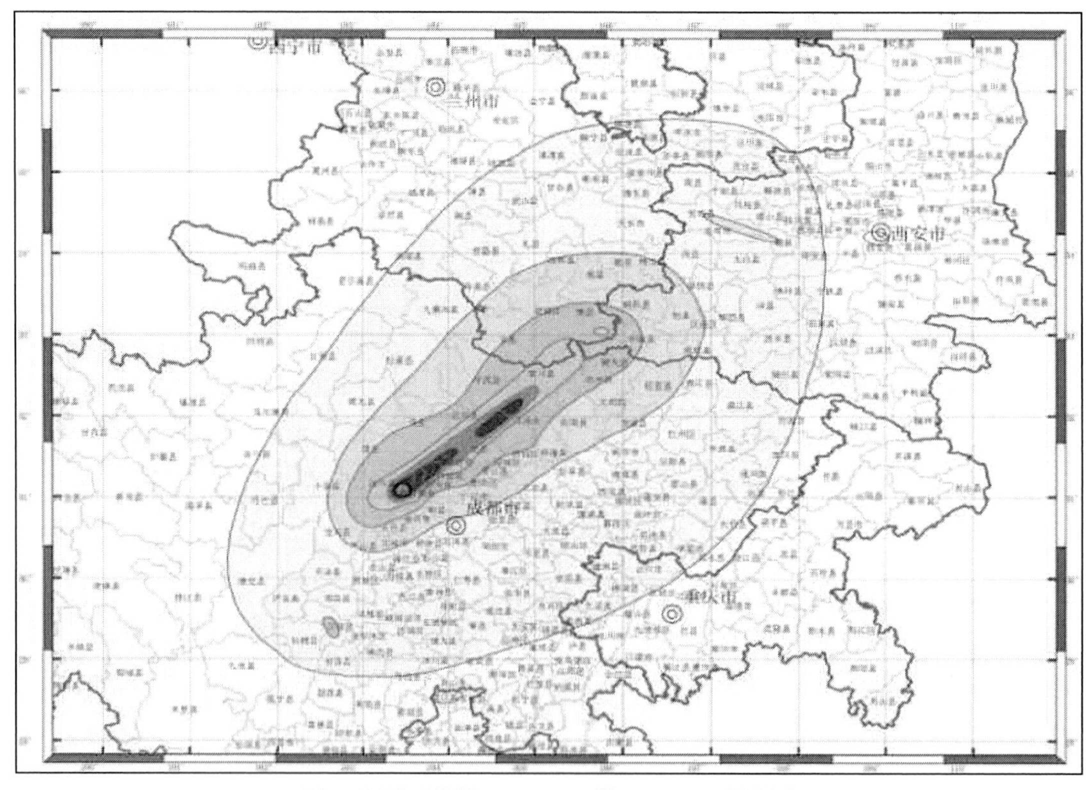

Fig. 2.20. MMI contours (Courtesy of IEM)

2.4 Landslide Failures

Landslides were prevalent wherever ground motions exceeded PGA = 0.35 g or so. Based on the field observations from the ASCE team, about 20 to 50 percent of the surface of the mountainous areas where PGA was > 0.35 g and slope was greater than 40 degrees had slid. Visual inspection of the mountainsides that did not slide did not suggest any obvious reasons (bedding, soil type, tree cover, or such) why one slope slide and the other did not.

Available statistics show 12,967 landslides (including rock falls and deep seated slides) and 501 debris flows situated as shown in Table 2.6. The number of debris flows is likely to increase over time with continued rainfall.

Figure 2.22 shows the fans of landslides entering a river. Along with high turbidity impacts, there were many landslide-formed lakes. These landslide-caused lakes (34 reported by Chinese authorities) in some places inundated low-lying areas and created a substantial flood hazard to downstream locations. The Chinese PLA worked diligently to open up waterways before landslide-made lakes became so dangerous that they failed the landslide-formed dams and created downstream floods.

The lake formed near Tangjiashan was the largest of the landside-created lakes. More than 100,000 people in Mianyang City were evacuated.

Figure 2.23 shows the dust created by the landslides within a minute or two of the earthquake.

Figure 2.24 shows a road cut into a hillside that was destroyed by a landslide. This was a common occurrence; essentially all roads leading from the Sichuan plain towards the Tibetan plateau, subjected to PGA > 0.30 g, were cut by such failures.

2.5 Surface Faulting

Figure 2.26 shows about 10 ft. of uplift along a concrete road.

Figure 2.27 shows 10 ft. of vertical upward thrust and 12 ft. of horizontal offset right at a road in Shenxigou Village. There was evidence of 1 ft. of offset at a few other locations within about 300 ft. of this location. Note that the offsets shown were observed in late October 2008; other investigators reported offset at 4.5 m vertical and 4 m horizontal at this site, but we would have to interpret those larger values as including the effects of subsidiary and secondary offsets.

About 6 km away also in Hongkou vicinity, Prof. Manchao He lead us to a surface fault shown in Figure 2.28. This faulting had a maximum vertical uplift of about 2 m with a length of about 30 m. It went through a house that collapsed (Fig. 2.29).

Table 2.6 Landslide Statistics

County	Number of landslides
Dujiangyan	10
Pengzhou	30
Shifang	9
Mianzhu	7
Mao Xian	61
Wenchuan	66
Li Xian	134
Heishui	49
Beichuan	24
An Xian	33
Pingwu	68

Fig. 2.21. Landslides (80 percent slope failure) (Courtesy IEM)

Fig. 2.22. Landslides with debris fans entering a river (Courtesy IEM)

Fig. 2.23. Landslide dust (Courtesy IEM)

Fig. 2.24. Failed roadway due to landslide

Fig. 2.25. A car crushed by a fallen rock

Fig. 2.26. Surface faulting through road at Gaoyan Village

Fig. 2.27. Surface faulting through road at Shenxigou Village

Fig. 2.28 Surface fault in Hongkou, Prof. He pointing at the smooth surface resulted from the uplift.

Fig. 2.29 This house was about 30 m from the surface fault show in Fig. 2.28. The stove at the back of the house was the only recognizable appliance.

Table 2.7. Site Classification in Chinese Code GB50011

Equivalent shear wave velocity v_{se} (m/s)	Site class			
	I	II	III	IV
$v_{se}>500$	0m			
$500\geq v_{se}>250$	<5m	≥5m		
$250\geq v_{se}>140$	<3m	3-50m	>50m	
$v_{se}\leq 140$	<3m	3-15m	>15-80m	>80m

2.6 Site Classification in Chinese Code GB50011

Construction sites are divided into four classes according to the equivalent shear wave velocity for surface soil layers and the overlaying thickness at the site (Table 2.7).

The overlaying thickness of the site shall be defined as follows:

(a) In general case, the overlaying thickness of the site is the distance from the ground level to the top of the soil layer where the shear wave velocity is more than 500 m/s.

(b) The overlaying thickness is the distance from the ground level to the top of the soil layer without 5 m below ground level, where the shear wave velocity is more than 2.5 times of that of the adjacent upper soil layer, and the shear wave velocity of underlying soil is not less than 400 m/s.

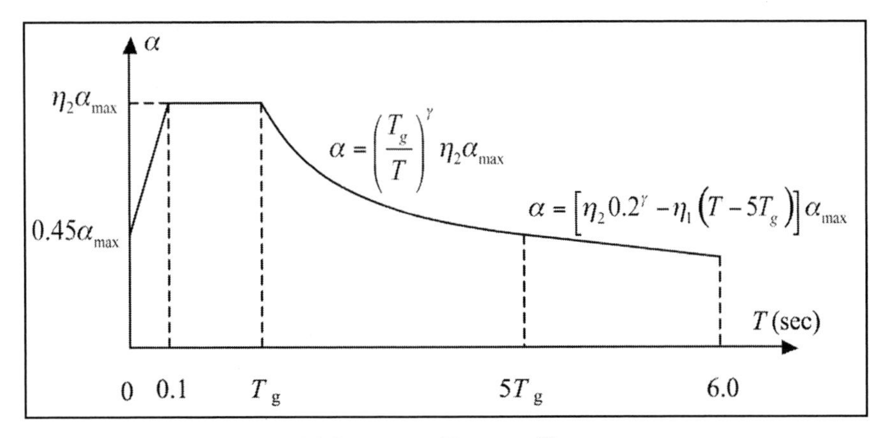

Fig. 2.30 Seismic effect coefficient curve

(c) The boulder with shear wave velocity more than 500 m/s, and the lens shall be regarded as surrounding soils.

(d) The volcanic rock interlayer shall be regarded as rigid body, and its thickness shall be deducted from the overlaying thickness.

The equivalent shear wave velocity shall be computed from the following equations:

$$\upsilon_{se} = d_0 / t \tag{1}$$

$$t = \sum_{i=1}^{n} (d_i / \upsilon_{si}) \tag{2}$$

where
d_0 = calculation depth, equals to the less value between 20m and overlaying thickness
t = shear wave travel time from ground level to calculation depth
d_i = the thickness of any soil layer i (between 0 and calculation depth)
v_{si} = the shear wave velocity of any soil layer i
n = the number of soil layer (between 0 and calculation depth)

2.7 Design Response Spectra in Chinese Code

Seismic effect coefficient of a building structure shall be determined from Figure 2.30 based on the class of design earthquake, site class, and natural period of a structure.

α = seismic effect coefficient, the value of α is the maximal absolute acceleration of mass relative to the acceleration of gravity;
$\alpha_{\max}$ = maximum value of seismic effect coefficient (Table 2.8);
T = nature vibration period of structure;
T_g = characteristic period taken from Table 2.9;
η_1 = decreasing slope adjustment coefficient of linear falling segment. If $\eta_1 < 0$, then $\eta_1 = 0$;
η_2 = damping adjustment coefficient. If $\eta_2 < 0.55$, then $\eta_2 = 0.55$;
γ = attenuation index of curvilinear falling segment.

The damping ratio of structures, ζ, should equal to 0.05 except special provisions, and $\gamma = 0.9$, $\eta_1 = 0.02$, $\eta_2 = 1.0$. For the structures having a period, T, more than 6.0s, the special study of seismic effect coefficient shall be performed.

If the damping ratio of structures is not equal to 0.05, γ, η_1 and η_2 shall be determined by the following equations:

$$\gamma = 0.9 + \frac{0.05 - \zeta}{0.5 + 5\zeta} \tag{7}$$

$$\eta_1 = 0.02 + (0.05 - \zeta)/8 \tag{8}$$

$$\eta_2 = 1 + \frac{0.05 - \zeta}{0.06 + 1.7\zeta} \tag{9}$$

The maximum value of horizontal seismic effect coefficient shall be taken from Table 2.8.

Table 2.8 Maximum Value of Horizontal Seismic Effect Coefficient (α_{max})

Intensity	6	7	8	9
Frequent earthquake	0.04	0.08(0.12)	0.16(0.24)	0.32
Rare earthquake	—	0.50(0.72)	0.90(1.20)	1.40

***Note**: The values in bracket are used in the region as basic design ground acceleration of 0.15g and 0.30g, respectively.*

Table 2.9 Characteristic Period Value(s)

Classification of design earthquake	Site category			
	I	II	III	IV
Group 1	0.25	0.35	0.45	0.65
Group 2	0.30	0.40	0.55	0.75
Group 3	0.35	0.45	0.65	0.90

***Note**: When rare earthquake intensities of 8 and 9 are evaluated, 0.05s shall be added to T_g.*

2.8 Acknowledgement

The authors would like to thank Zifa Wang, Ph.D., of the Institute of Engineering Mechanics (IEM) and a member of China Earthquake Administration (CEA) for providing valuable geotechnical information and photos for this chapter. We are also indebted to Professor Manchao He for showing the study team members the location of the fault uplift in Hongkou as well as providing a debriefing of the local geology prior departing to Chengdu, Sichuan.

Unless otherwise referenced in the table or figure captions/sources, all figures and tables are owned by the members of ASCE/TCLEE teams.

3 TRANSPORTATION SYSTEMS

EXECUTIVE SUMMARY

The Wenchuan Earthquake occurred in a mountainous region where landslides caused damage to roads and railway lines. In the days following the earthquake, aftershocks, rain, mud flows, and debris dams made matters worse. Bridges in the area hadn't been adequately designed for earthquakes, resulting in the loss of many river crossings. Surface faulting also damaged roads and bridges. Approximately 30,000 miles of roads and railways, 3,000 bridges, 100 tunnels, and many miles of retaining structures were damaged by the earthquake. Losses to the transportation sector exceeded $10 billion.

On the day of the earthquake, 31 passenger trains and 149 cargo trains were stranded on lines linking Chengdu with the rest of the country. There were many landslides, bridge collapses, and other damage along rail tracks, and 34 railway stations on the Baoji-Chengdu Railway lost power due to the earthquake. A cargo train on the Baoji-Chengdu Railway caught fire in a tunnel near Huixian County in Gansu Province as the tunnel collapsed. The Railways Ministry dispatched rescue teams to the train and sent repair teams to check railway facilities in the region. All trains running in the affected area were ordered to halt in open areas, and passenger trains heading for quake-hit areas were ordered to turn back. By midnight on the day of the earthquake, a number of trains from Chengdu heading for Kunming, Wuchang, and Nanchang had resumed operations, although about 10,000 passengers were still stranded at the Chengdu Railway Station.

Major highways and expressways in Sichuan were also closed, according to Mr. Mengyong Weng of the Ministry of Transport. Transport along the expressway linking Chengdu to Mianyang was halted. He also reported that roads linking Wenchuan to the city of Dujiangyan were damaged, blocking disaster relief teams within the city. Landslides also struck several highways in neighboring Shaanxi Province, while the national highway linking it with Sichuan Province remained in operation. The Ministry of Transport was immediately engaged in repairing damaged roads to ensure that the quake-hit areas had access to goods for disaster relief. The ministry asked neighboring provinces to repair roads leading to Sichuan.

Li Jiaxiang, acting director of the General Administration of Civil Aviation of China (CAAC), went to Sichuan to direct disaster relief work after the earthquake. Chengdu Shuangliu International Airport was temporarily closed affecting 169 inbound and 108 outbound flights. Some facilities were damaged, and the control tower and regional radar control were evacuated. The airport reopened on the evening of May 12, offering limited service as the airport became a staging area for relief operations. Jiaxiang was accompanied by top officials from Air China, who helped with transportation of disaster relief goods. All airports in Sichuan were reopened by the end of the day.

After the earthquake, an effort was made to repair roads and railroads as quickly as possible. Fourteen days later, all roads into Chengdu were repaired. By June 14, roads to 248 of the 254 towns in the affected area were repaired. This was usually a single, unpaved lane with traffic control. Railways took a little longer because they needed a smoother alignment and electric power for their operation. Repairs will continue for many months and years. At the time of the earthquake, roads and bridges were being built on an accelerated pace in this region. Hopefully, lessons will be learned, and the new alignments will be less vulnerable to strong shaking, landslides, and the large fault offsets that occur in this region.

Relief supplies were delivered in two ways. There are five airports around Chengdu including two People's Liberation Army (PLA) (military) airports. Relief supplies were shipped to these airports and then trucked into the areas of damage. MI-26 helicopters were used where road damage blocked passage. This included the delivery of graders and other earthmoving equipment for removing landslides and road repair.

3.1 Performance of Roads

An investigation of road damage was performed by 11 teams organized by China's Ministry of Transport with support from the Sichuan Province Highway Department following the earthquake. An aerial survey was also completed to help the ministry identify areas of major road damage (Zhang 2009). They found 21 expressways, 15 provincial highways, and 2,795 rural roads were damaged. There was a total of 28,000 km of road damage including 200 km of expressway damage, 3,849 km of provincial highway damage, and 23,800 km of rural road damage (Fig. 3.1).

Fig. 3.1. Immediately after the earthquake, every road into the earthquake region was closed. Fourteen days after the earthquake, the roads into Chengdu had been re-opened, By June 14, roads into 248 of the 254 towns were repaired, although many still had not been repaved (Courtesy of Ministry of Transport of China)

The earthquake had a very large magnitude, which increased the amplitude and duration of ground shaking, enlarged the area impacted, and resulted in a great deal of road damage. Additionally, the earthquake occurred in the mountains, which increased the number of hazards, caused more road damage, and made repairs more difficult. Several million people lived in the area of heavy shaking. There were about 90,000 casualties and 400,000 injuries, and the road damage made the emergency response more difficult.

There was road damage due to rockfalls, avalanches, landslides (Zang 2009), debris flow (Tang 2009), liquefaction with lateral spreading, quake lakes, surface rupture, and collocation effects. Heavy rains occurred after the earthquake, causing more slides, making damaged roads more unstable, and creating miserable conditions the population. Many of the roads are switchbacks with steep grades, which made travel in wet weather more dangerous.

The team traveled with members of the Sichuan Province and China Ministry Road Departments, who were reluctant for us to travel in such dangerous conditions a month after the earthquake. The roads were rapidly regraded (usually as single lane unpaved and unflagged roadways), but it will be sometime in 2010 before the highway system returns to its pre-earthquake condition.

3.2 Road Damage Due to Rockfalls, Landslides, and Mud Flows

The earthquake occurred in the mountains west of Chengdu (called the Longmen Shan or Dragon's Gate Mountains). These mountains are steep, continually eroding, and subject to landslides even when there are no earthquakes. Streams cut deeply into the mountains, removing the toe of the slope, which resulted in unstable soil and rock masses (Fig. 3.2). During the May 12 earthquake, there were thousands of avalanches and landslides (Fig. 3.3). Roadway cuts also made the mountainsides more unstable.

Rockfalls

During the May 12 earthquake, thousands of large rocks that had precariously balanced on mountainsides fell onto roads killing drivers, damaging the roadbed, and closing the roads until heavy equipment could be sent to clear the site (Fig. 3.3). This occurred at hundreds of locations through the mountains. Some of the rocks weighed more than 3000 tons and were difficult to move. Fortunately, the PLA had several Russian-made MI-26 helicopters capable of a 20-ton payload and was able to bring earth moving equipment to hundreds of sites. The damaged roads prevented food and water from getting to many mountain communities. The residents of many villages had to hike (or be carried) out of the mountains to survive.

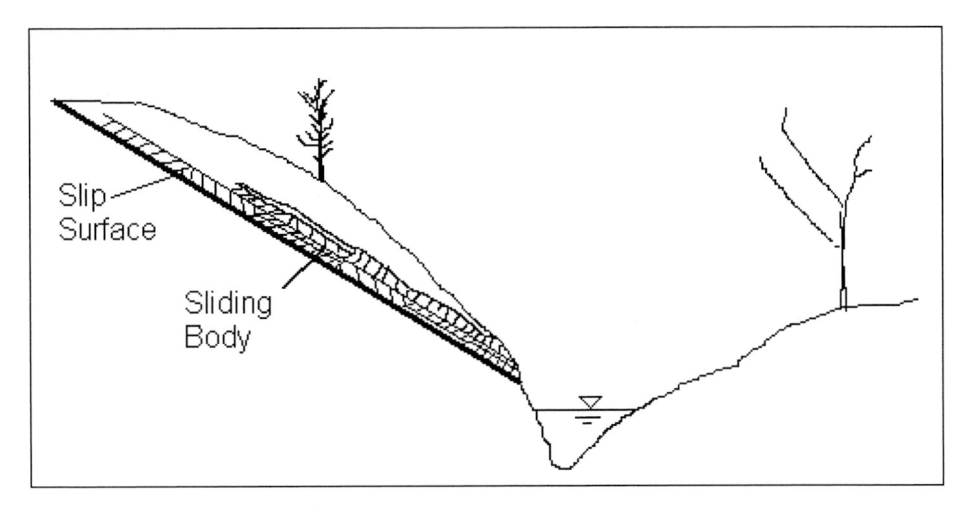

Fig. 3.2. Removal of toe of slope by fast-running mountain streams

Fig. 3.3. Rock slide in Gaochuan, Anxian County

Landslides

Thousands of landslides denuded mountainsides and damaged roads. Sometimes the slope above a road slide down and covered the roadway with tons of material to be cleared away (Fig. 3.4). Equally common, the slide carried the roadway downhill (Fig. 3.5). These landslides not only affected roads, bridges were also vulnerable (Fig. 3.6).

Fig. 3.4. Landslides blocked many roads

Fig. 3.5. A landslide carries a roadbed downhill near Pengzhou

Fig. 3.6. Landslide destroyed a bridge on Duwen Road

Mud Flows

Mud flows occurred during the earthquake and in the following weeks and months. As a result, many quake lakes were formed by mud-flow created dams during and after the earthquake. Rocks, soil, and other debris loosened by the strong ground shaking during was prone to form mud flows. Heavy rainfall on the evening of August 7, 2008, caused a mud flow in Pengzhou that lasted about 40 minutes. The mud flow had ceased by the following day, but the debris—boulders up to 1.5 m in diameter, tree trunks up to 30 cm in diameter, and a sand/silt clay matrix with a very high moisture content—was still in a very fluid state with low bearing capacity, as evidenced by individuals sinking to their knees as they traversed it. This mud flow covered the roads and prevented transportation into and out of the area (Fig. 3.7 and 3.8).

Fig. 3.7. Mud flow in Pengzhou

Fig. 3.8. Mud flows covered several roads

3.3 Road Damage From Liquefaction, Surface Rupture, Quake Lakes, and Collocation

Liquefaction

Some of the damage, especially along the rivers, looked like liquefaction but on closer inspection turned out to be due to surface rupture. While there probably were roads damaged by liquefaction, we saw only liquefaction damage to railways during our investigation.

Surface Rupture

The Longmen Shan thrust belt consists of three faults: the frontal fault (Dujiangyan-Jiangyu), the central fault (Yingxiu-Beichuan), and the back fault (Wenchuan-Maoxian) (Fig. 3.9). The total rupture length was about 300 km. Surface rupture was observed along the central fault and the frontal fault (Fig. 3.10). No surface rupture was observed along the back fault.

The observed surface rupture along the Yingxiu-Beichuan segment was about 210 km, and the vertical offset varied from about 1.0 to 2.0 m near its southern end to 5.0 m in its center (south of Beichuan) to 1.0 to 2.0 m near its northern end. Figure 3.10 shows an example from Yingxiu Town near the southern end, with a vertical offset of about 2 m and right-lateral offset of about 0.5 m (location a in Figure 3.9). Collapsed buildings can be seen on the hanging wall in Figure 3.10. Location b in Figure 3.9 is near the segment's center. Pintong Town of Pinwu County (location c in Figure 3.9) is near the northern end of the segment. A surface rupture of about 70 km was also observed between Wafeng Town of Dujiangyan and Hanwang Town of Mianzhu along the Dujiangyan-Jiangyu segment (location d in Fig. 3.9).

Vertical and right-lateral offsets of about 1.0 and 0.5 m, respectively, were observed along the Dujiangyan-Jiangyu segment.

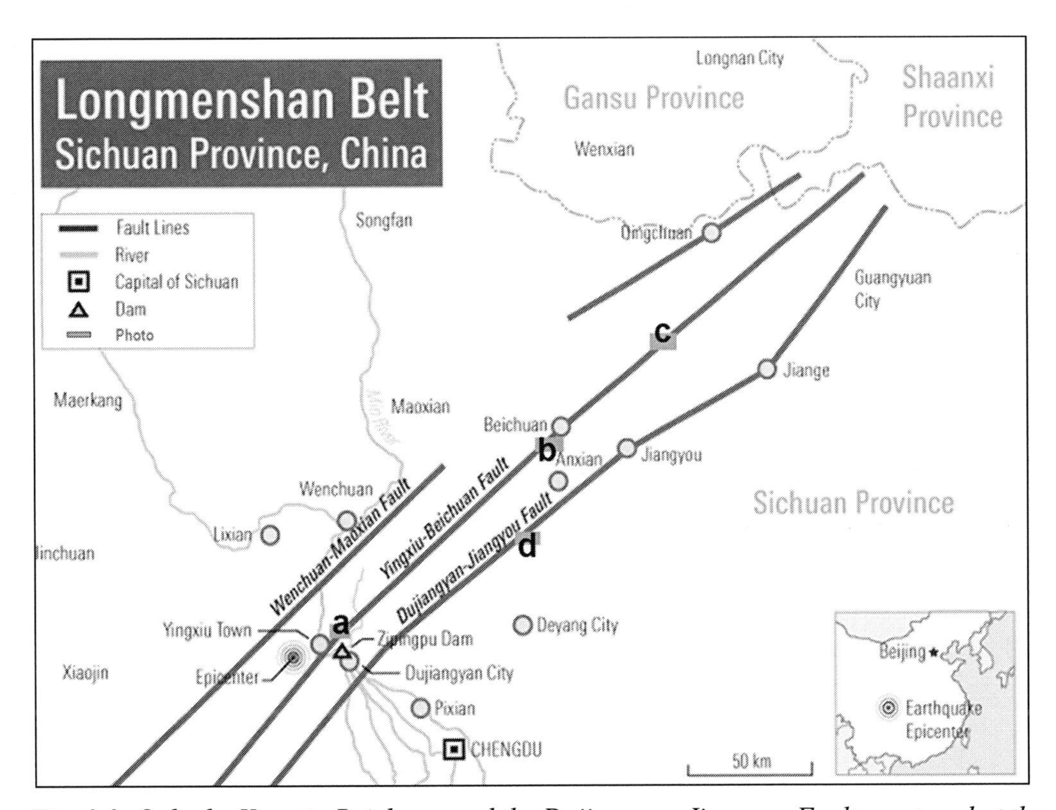

Fig. 3.9. Only the Yingxiu-Beichuan and the Dujiangyan-Jiangyou Faults ruptured at the surface (Courtesy USGS)

Fig. 3.9a Yingxiu Town near the southern end, with a vertical offset of about 2 m and right-lateral offset of about 0.5 m

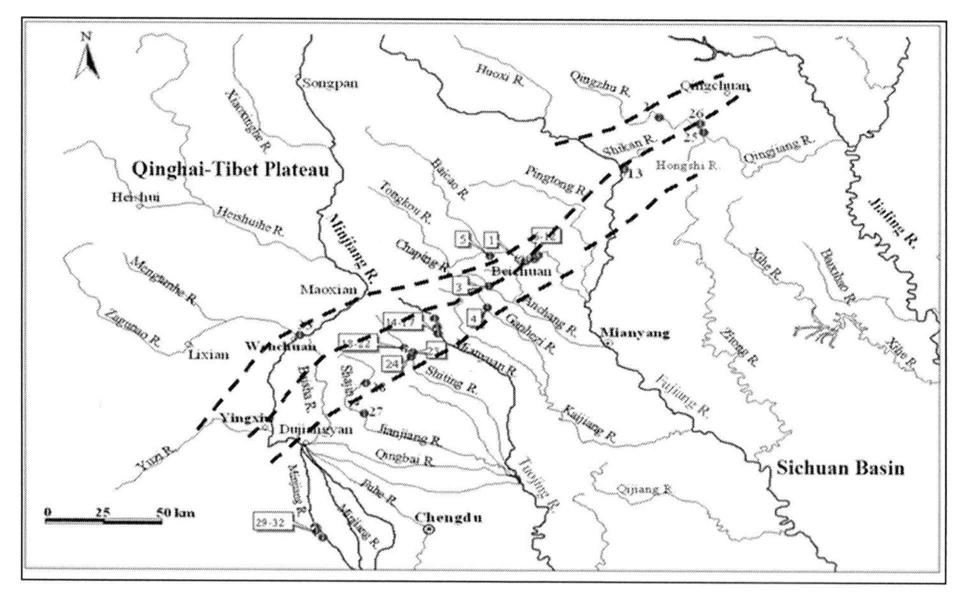

Fig. 3.10. Locations of 33 quake lakes formed by landslides into rivers. (Courtesy of IEM)

Quake Lakes

Field surveys identified 33 dammed lakes in Beichuan (9), Qingchuan (4), Anxian (6), Pingwu (1), Songpan (2), Mianzhu (3), Shifang (4), Pengzhou (2), and Maoxian (2).

Tangjiashan Lake was formed when a landslide dammed the Jian River (Fig. 3.11). Its name comes from the nearby mountain Tangjiashan. On May 24, 2008 the water level reached a depth of 23 m (75 ft.) covering roads and washing away several bridges (Fig. 3.12). Soldiers used digging equipment, explosives, and missiles to blast channels in the dam. The flow from the resulting channel caused flooding in the evacuated town of Beichuan and overtopped several dams.

Fig. 3.11. Tangjiashan "quake lake" formed by landslide and mud blocking the Jianjiang River (31.643 Degrees, 104.429 Degrees) (Courtesy IEM)

Fig. 3.12. Quake lake closed roads and damaged bridges

Fig. 3.13. Colocation damage, such as the retaining wall and telephone pole blocking access to the roadway, is very common after earthquakes

Colocation Damage

Colocation damage is caused by the failure of a structure or a lifeline affecting an adjacent lifeline, such as a retaining wall, utility line, or building failing closing a road (Fig. 3.13 and 3.14).

Fig. 3.14. Colocation damage due to collapses buildings blocking roads in Yingxiu

Most of the damaged roads were in the mountains. Although there was too much damage to describe completely, some of the most damaged routes are illustrated in the following section.

3.4 Road damage between Dujiangyan and Yingxiu

At the time of the earthquake, the only road between Dujiangyan and Yingxiu was Provincial Highway 213. The construction of a 26-km expressway was almost completed when the earthquake occurred. The existing highway was almost completely destroyed, and the Miaoziping Bridge (across the Zipingpu Reservoir) on the new expressway partially collapsed, 295 m of the Longxi tunnel was damaged, and 1,931 m of Zipingpu tunnel was damaged. There was damage of 2.2 billion Yuan ($323.5 million) to the new expressway.

When the investigation team toured the area in July, existing Route 213 had been regraded, but it was still an unpaved one lane road that had no traffic control and was continually washing out due to heavy rains. There was still evidence of rockfalls and landslides everywhere (Fig. 3.15 and 3.16). As we drove into Yingxiu Village, there was so much rubble from damaged buildings that it felt like driving through a huge maze. Many of the bridges had been damaged, but the road remained opened because either the ground was graded alongside of the bridge or the bridge was held up by shoring.

Fig. 3.15. Fallen rocks on Route 213 after the May 12 earthquake

Fig. 3.16. Landslide and ground shaking damage at Zhonghi Bridge near Yingxiu

Reconstruction of the new expressway started in June 2008, and it was opened to traffic on May 13, 2009. The finished expressway has two lanes in each direction and will be used to help rebuild the area of severe damage.

3.5 Road Damage Between Yingxiu and Wenchuan

In the 20-km-long highway from Yingxiu to Wenchuan, 340 landslides and rockfalls occurred, and the highway was blocked for six months (Fig. 3.17). At the time of the earthquake, traffic density was at its yearly peak due to the sightseeing traffic to and from the Jiuzhaigou Valley UNESCO World Heritage site. According to the numbers of destroyed buses and cars, it is estimated that there were more than 1,000 deaths on the road. The damage was not caused by surface faulting because the adjacent Wenchuan Maoxian fault didn't rupture to the surface during this earthquake.

3.6 Road Damage Between Wenchuan and Qingchuan

There were many rockfalls, landslides, mudslides, debris dams with flooding, and surface faults on the road between Wenchuan and Qingchuan (Fig. 3.18).

Fig. 3.17. Landslide on Route 213 between Yingxiu and Wenchuan

Fig. 3.18. Surface fault damaged a road between Beichuan and Nanba

Fig. 3.19. Typical retaining wall failure along roads in Sichuan

3.7 Retaining Walls

It is difficult to retain downward sloping material excited by ground motion greater than a few tenths of a g. Moreover, most of the retaining structures were made of unreinforced concrete filled with cobbles and were barely adequate for non-seismic loads (Fig. 3.19). Surprisingly, many of these structures we saw very close to the fault were undamaged (Fig. 3.20). This illustrates the spatial variability of the ground motion rather than indicating the remarkable strength of the unreinforced stone masonry retaining walls.

Fig. 3.20. Same type of retaining wall as Fig.3.19 but without damage

Fig. 3.21. Failure of MSE wall supporting a roadway

Mechanically stabilized earth (MSE) retaining systems were inadequately anchored into the surrounding material and failed (Fig. 3.21). We have seen good seismic performance from MSE walls during previous earthquakes when they were adequately designed.

We also saw many rock anchors that did an excellent job holding back dam abutments and other essential lifelines (Fig. 3.22).

Fig. 3.22. Rock bolts did a good job of securing Zipingpu Dam abutments

3.8 Bridges

We observed many bridge damage throughout the field trip, including damage to new bridges, bridges under construction, and existing bridges. Many arch bridges remained undamaged.

According to the *Specifications of Seismic Design for Bridges*, JTJ 004-89, China, earthquake design loads for bridges are computed as follows:

Where W = dead weight of bridge, C_i = importance factor, C_z is a response modification factor, beta is a elastic seismic response coefficient, K_h is an acceleration coefficient.

- C_i = 1.7 (express highway), 1.3 (arterial highway), 1.0 (secondary road), 0.6 (3rd or 4th level highway)
- K_h = 0.1g (intensity VII), 0.2g (intensity VIII), 0.3g (intensity IX)
- Beta = elastic response spectra shape, which is soil profile dependent. For all soils, beta is capped at 2.25 in the high frequency range. For soil type I (rock), beta = 2.25 x 0.2/T where T = structure period > 0.2 sec. For soil type II (firm soil), beta = 2.25 x (0.3/T)**0.98 where T = structure period > 0.3 sec. For soil type III (deep soil), beta = 2.25 x (0.45/T)**0.95 where T = structure period > 0.45 sec. For soil type IV (deep soft soil), beta = 2.25 x (0.7/T)**0.9 where T = structure period > 0.7 sec.
- C_z is a factor to represent ductility (see Table 4.1).

Table 3.1. Response Modification Factor Cz

Bridge Type	Pier Height / Type	H < 10 m	10 ≤ H < 20	20 ≤ H < 30
Girder	Flexible	0.30	0.33	0.35
Girder	Massive	0.20	0.25	0.30
Girder	Pile Shaft pier	0.25	0.30	0.35
Arch		0.35	0.35	0.35

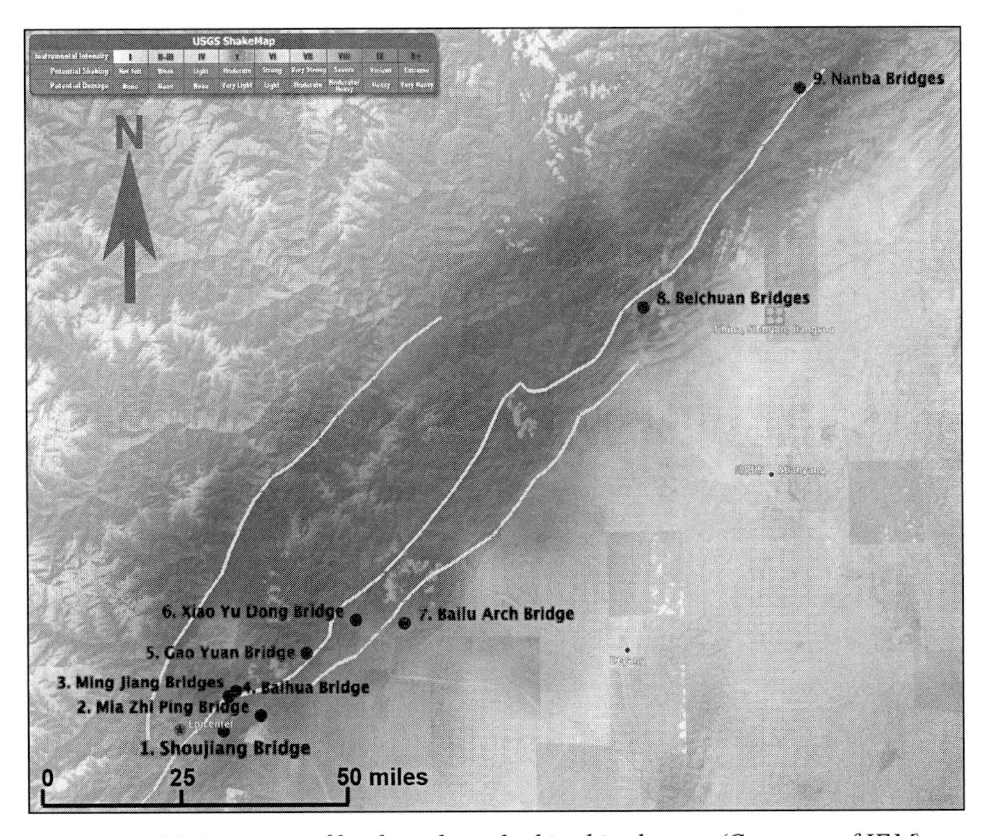

Fig. 3.23. Location of bridges described in this chapter (Courtesy of IEM)

For example, assume the Gaoyuan bridge (Fig. 6-7) is in an intensity VII zone with firm soil and T = 0.6 seconds. Also assume this is a secondary road (Ci = 1.0), Kh = 0.1g, Beta = 1.14. Cz = 0.25. Then, V = (1.0) ≅ (0.25) ≅ (0.1) ≅ (1.14) ≅ W, or V =

0.0285 W. Clearly, this low level of design force (under 3 percent of weight) would allow the designer to select girders that rest on the bent caps and rely on friction for lateral resistance only, as the coefficient of concrete on just about anything is much higher than 0.03. As the actual ground motions (PGA) at this bridge likely ranged from 0.4 g to 0.6g , gross failure would be expected. It is unlikely that the bridge would have fared any better had the bridge been designed to Intensity IX (PGA = 0.3

g, or V=0.09W) because the other factors in the code seismic force formulation are too low or the assigned ductility capacity is too high for the style of construction.

The damage to bridges ranged from cracks to complete collapse. Figure 3.23 shows the location of bridges described in this chapter.

Shoujiang Bridge (30.9800°N, 103.4605°E)

The Shoujiang Bridge is as an eight-span, 280-m-long, T-girder bridge supported on very tall two-column towers and pier walls. The tallest piers were approximately 60 m high, the approach spans were 30 m long, and the four main spans were 40 m long. Both the towers and the pierwalls were supported on pile foundations (see Fig. 3.24). The north span almost fell off the north tower and was supported by a temporary steel tower (Fig. 3.25). A Bailey bridge had been launched over the north span to carry the traffic. The north approach embankment showed evidence of ground failure, and there was a shear failure of the north abutment (Fig. 3.26). It appeared that too much movement of this very tall structure pushed the girders of the first span into the approach and almost unseated it from the tower. It is unclear if the hazard was faulting, a landslide, or ground shaking. The most serious vulnerability was the short seat that could not accommodate the resulting movement at Bent 2.

Miaoziping Bridge (N31.0208°, E103.54461°)

The bridge over Zipingpu reservoir had one collapsed span and many other spans with displaced girders. The construction of this bridge started in 2007, and it was scheduled to open in June 2008. The main span (right side of Fig. 3.27) is a continuous rigid frame; the remaining approach spans are a simply supported girder-type. Figures 3.28 and 3.29 provide details of the bridge with the missing span.

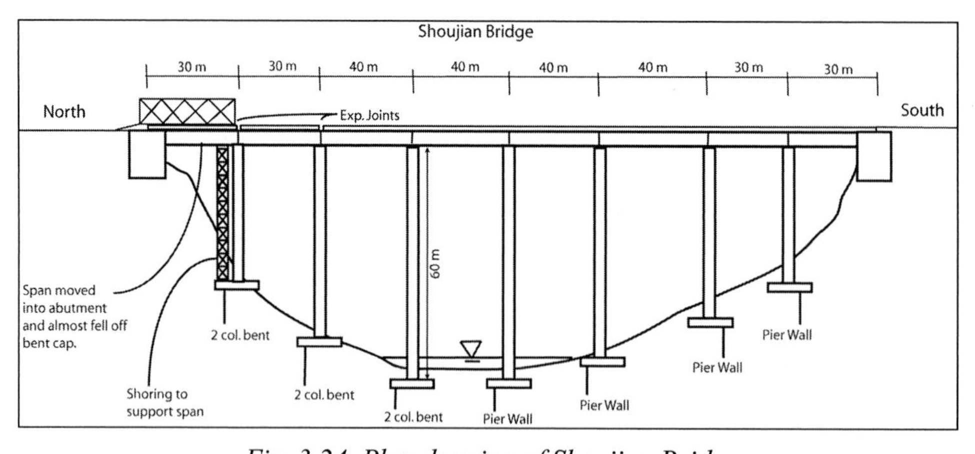

Fig. 3.24. Plan drawing of Shoujian Bridge

Fig. 3.25. Photos of dropped span at Bent 2 of Shoujian Bridge

Fig. 3.26. Photos of damage at the north abutment

Figures 3.30 and 3.31 show some of the other spans on this bridge. A few lateral shear keys sustained larger cracks. This type of shear key design was used for many bridges in the area and was also observed in bridges in Beijing. It is apparent that it has insufficient lateral strength (or ductility) to withstand high loads from the bridge deck associated with strong inertial shaking.

The team visited the Miaoziping Bridge a second time on October 17, 2008 (158 days post-earthquake). There was still considerable heavy equipment on the bridge making repairs (Fig. 3.32 and 3.33). The missing span was replaced. Note that the conduit along the side of the bridge deck carries electric power lines. Figure 3.34 provides the plan and elevation views of the bridge; note that the bridge enters tunnels at both ends.

This bridge is part of the new expressway that was being built from Dujiangyan to Wenchuan at the time of the earthquake (described in the section on road damage). It was reported that the expressway had been completely repaired on May 13, 2009, along with this bridge and several tunnels.

Fig. 3.27. Miaoziping Bridge, taken in October 2008

Fig. 3.28. Miaoziping Bridge with Missing Span (13th span), many other spans shifted (taken in July 2009)

Fig. 3.29. Close-up of the Miaoziping Bridge with missing span shown in Fig. 3.28

In May 2009, a small TCLEE team attended a symposium hosted by the Department of Sichuan Association for Science and Technology on lifeline earthquake loss reduction to support the re-building of the disaster areas. The team had an opportunity to meet with the Director Feng, whose organization was responsible for the repair and management of the accelerated reconstruction of the Miaoziping Bridge and the tunnels associated with the Duwen Expressway. The bridge and the tunnels were repaired on time and opened on the anniversary of the earthquake rendering access to Yingxiu and Wenchuan easier. Figure 3.35 shows the team on the repaired span.

Ming Jiang Bridge and Other Structures Near Yingxiu (N31.0443°, E103.475°)

The Ming Jiang Bridge at Yingxiu is located a few hundred meters north of the Beichuan Yingxiu fault and was very close to the epicenter of this earthquake. This area suffered intense ground shaking, landslide, and surface faulting damage (Fig. 3.36).

Fig. 3.30. Miaoziping Bridge with shear keys damaged

The Ming Jiang bridge was built in 2007. It is a T-girder structure with two-column bents that are supported on pile shafts and a span arrangement of 4-25m + 1-27m (Fig. 3.37). The Sichuan Province government built a Bailey bridge to carry vehicles over the its east span, which was damaged by a landslide. Several other structures in the vicinity were damaged, including an elevated viaduct, the Longxi Tunnel, and a levee, along with most of the village.

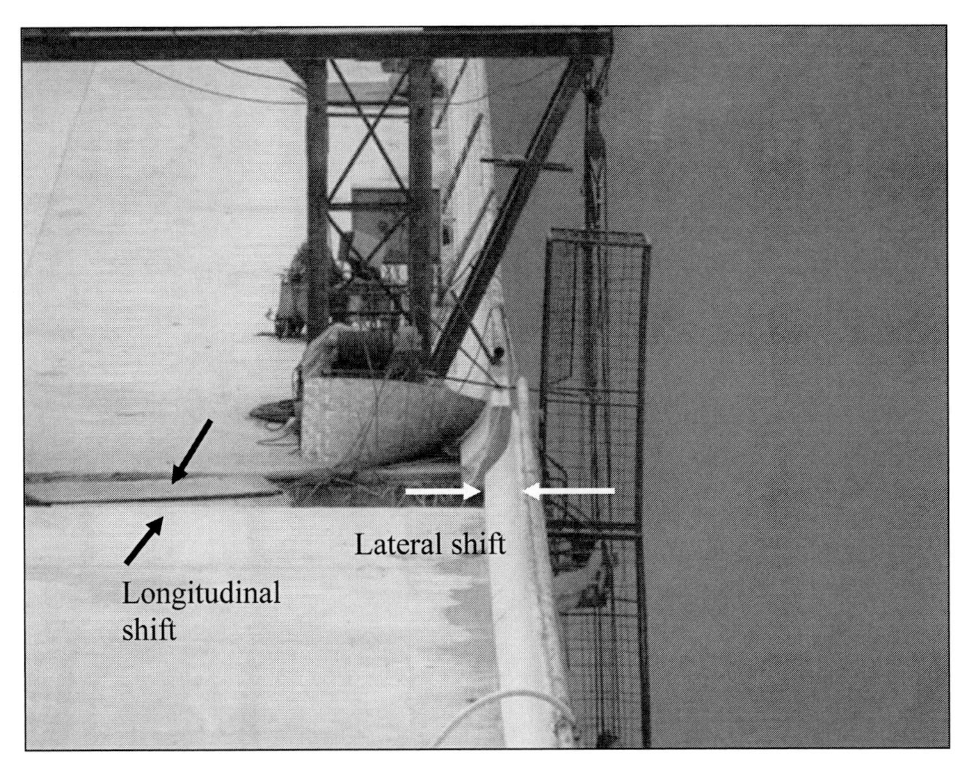

Fig. 3.31. Miaoziping Bridge with a permanent displacement of about 20 cm after the earthquake

Fig. 3.32. Miaoziping Bridge, heavy equipment on the bridge making repairs

Fig. 3.33. The conduit under the side of the bridge carries electric power

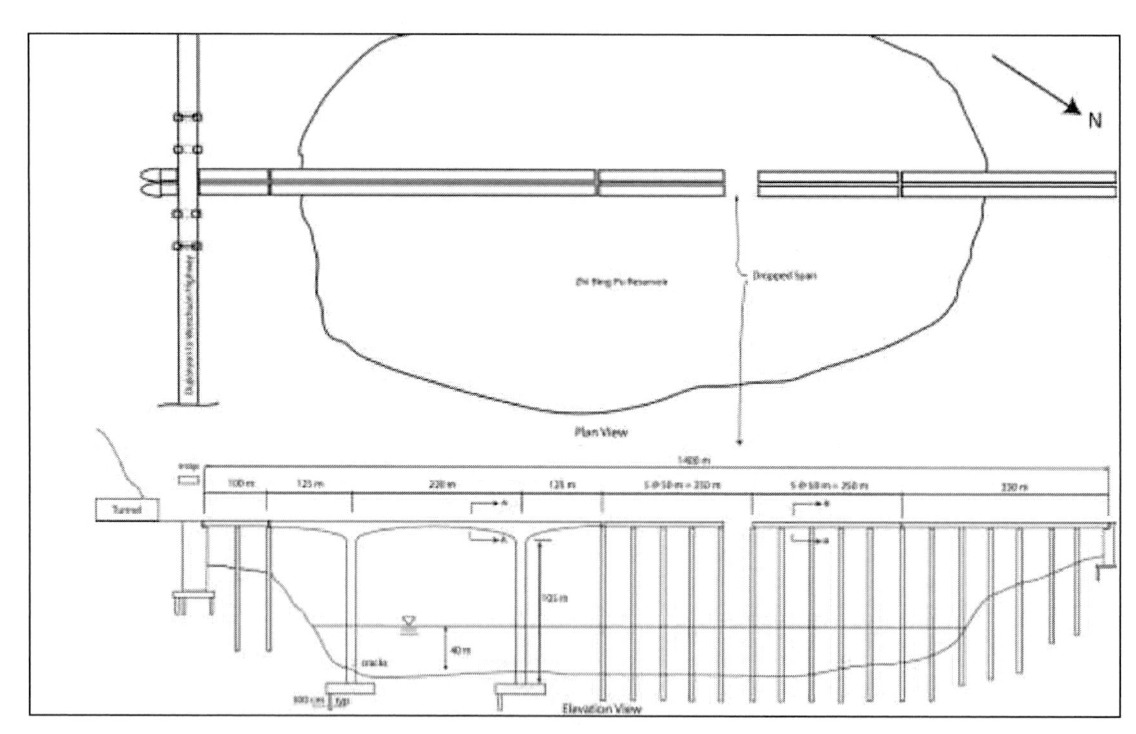

Fig. 3.34. Miaoziping Bridge plan and elevation views

Fig. 3.35. Repair team; third from the right is Director Feng, who lead his team to complete the re-construction on time

Figure 3.38 shows the remnant piers of a collapsed four-span bridge in Yingxiu. A nearby three-span bridge remained in service, despite moderate spalling damage at the shear keys, damaged bearings atop the piers, and moderate approach settlement damage at the abutments (Fig. 3.39 and 3.41). A newly built temporary water pipeline as well as new fiber optic cable and low-voltage power cable were supported on the bridge.

Baihua Bridge (N31.0443°, E103.4749°)

The Baihua Bridge is part of a Class 2 highway from Dujiangyan to Wenchuan. It was built in 2004 by the owner of a nearby hydroelectric plant to transport workers. Figure 3.42 is an aerial view of the damage. Three spans along with the piers collapsed.

As shown in Figure 3.43, the bridge is an 18-span, reinforced concrete structure with a total length of 450 m. The bridge superstructure was supported on two-column bents of varying heights as it climbs over the hilly terrain. The tallest bents have one or two struts to provide transverse restraint between the columns. The bridge has both straight and curved spans. For convenience, the bridge structure can be divided into six sections as summarized in Figure 3.43. The superstructure was a prestressed box girder with a drop-in T-girder span between Bent 9 and Bent 10. There were expansion joints at Bents 2, 6, 9, 10, and 14, and at the two-seat-type abutments. The bridge deck of the drop-in span just rested on the bent cap at its both ends.

Fig. 3.36. Aerial view of structural damage in Yingxiu Village (Deren Li, 2009)

During the earthquake, the more highly curved section of the bridge completely collapsed as illustrated in Figure 3.43. The rest of the bridge suffered varying degrees of damage, including shear cracks and failure at columns and struts, shear key failure, and bearing failure for Bents 3, 9, 15, and 18, respectively. At Bent 3, typical spalling and crack damage occurred between the strut and columns. At Bent 9 with expansion joints, the superstructure had significant transverse displacement, knocking off the shear key. At Bent 15, the bridge section was completely collapsed due likely to the shear and flexural failure of columns. At Bent 18, in addition to cracks between column and strut, significant spalling occurred underneath the bridge deck.

Fig. 3.37. Landslides, surface faulting, damaged Ming Jiang Bridge, and other structural damage in Yingxiu Village

Fig. 3.38. Collapsed four-span bridge at Yingxiu

Fig. 3.39. Damaged but in-service three-span bridge at Yingxiu

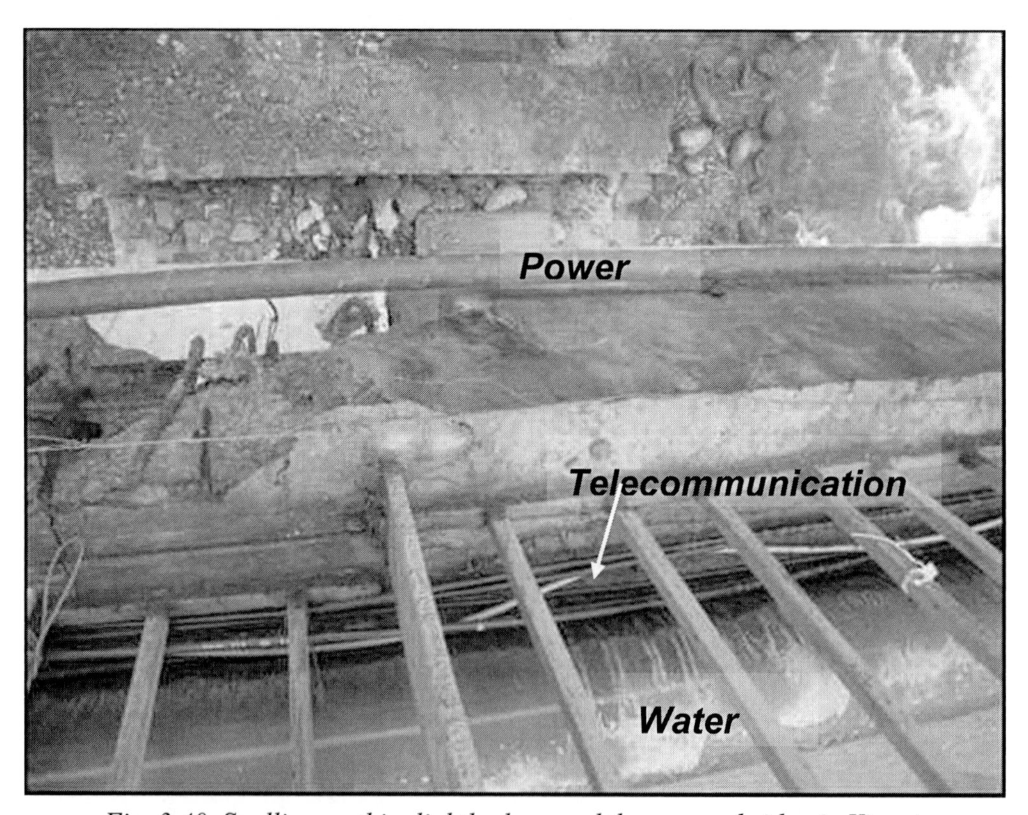

Fig. 3.40. Spalling on this slightly damaged three-span bridge in Yingxiu

Fig. 3.41. Temporarily rerouted utilities (power, telecommunication, and water) on the three-span bridge, Yingxiu

Figure 3.44 shows the failed base of a pier column, the transverse reinforcement seemed to undersized and widely spaced. The Ministry of Transportation told us that this bridge had been privately constructed to transport workers to a nearby power plant. Because the bridge collapse blocked access for vehicles, the bridge was dynamited and a road was graded alongside of it.

Fig. 3.42. Collapsed Baihua Bridge

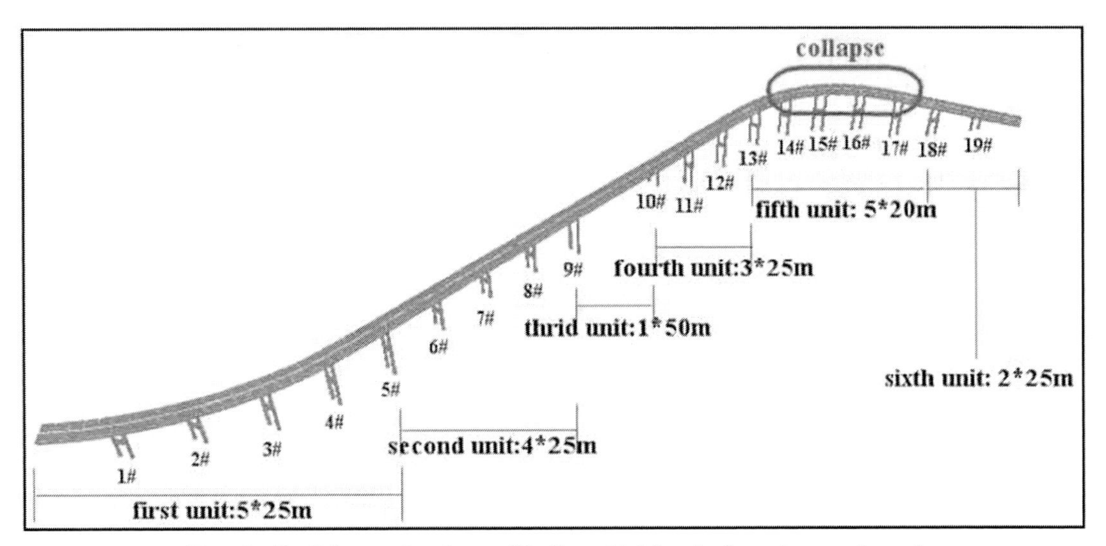

Fig. 3.43. Schematic view of Baihua Bridge before the earthquake

Fig. 3.44. Circular column with hinging (widely-spaced hoop steel) for Baihua Bridge

Gaoyuan Bridge (N31.106°, E103.637°)

It is speculated that the failure of the Gaoyuan Bridge was caused by a large velocity pulse that pushed the left abutment towards the right and soil compressive loading that punched out the unreinforced abutment side wall (seen on left side of Figure 3.45), resulting in sliding of the unrestricted concrete span girders towards the left. Some observers suggested the damage was due to fault offset, but we did not observe this as a likely failure mode. The unreinforced masonry damaged abutment is seen on the left of the photo (Fig. 3.45). The collapsed span reflects unseating due to excessive motion along the long axis of the bridge. We examined this span (Fig. 3.46) and found that every one of the concrete girders sits without any form of positive attachment to the pier caps. Other spans were shifted but did not move off the pier cap to fall. The concrete girders included no steel or any other positive attachment to the abutments or the bent caps. The small sign in Figure 3.47 under the remaining supported span reads, "do not dally under this bridge." It was sound advice, as the remaining girders would likely become unseated should a small aftershock occur. The side shear keys on this bridge showed evidence of substantial lateral shifting of the girders (Fig. 3.48), indicating that they were grossly underdesigned for the actual seismic lateral loads. Additionally, the keys were completely destroyed to the extent of having almost no residual capacity at the center pier cap (Fig. 3.48).

Fig. 3.45. Gaoyuan Bridge near Hongkou

Fig. 3.46. Failed span of the Gaoyuan four-span bridge (Hongkou)

Fig. 3.47. Gaoyuan four-span bridge (Hongkou)

Fig. 3.48. Close up view of Gaoyuan Bridge's failed span (Hongkou)

This bridge supported a series of low-voltage distribution cables, as can be seen hanging from the right side in Figure 3.48. While we could observe no clear break in these cables, a new power line had been erected, suggesting that the old sagging lines faulted.

Xiayudong Bridge (N31.1859°, E103.7667°)

The collapsed Xiayudong Bridge was constructed in 1998 (Fig. 3.49 and 3.50). The bridge is 187 m long and 23 m wide, with four arches and a main span of 40 m. The bridge is near a fault rupture zone. Figure 3.49 shows two damaged but standing spans; Figure 3.50 shows two adjacent collapsed spans.

Fig. 3.49. Damaged Xiaoyudong Bridge (Longmenshan town and Xiaoyudong town)

Fig. 3.50. Collapsed Xiaoyudong Bridge span (Longmenshan town and Xiaoyudong town)

Xiaoyudong (little fish hole) Bridge is a four-span reinforced concrete arch structure with long approaches, supported on two abutments and three intermediate bents as schematically shown in Figure 3.51. Each bent consists of one cap beam and two rectangular columns; it is supported on two drilled shaft foundations, each 2 m in diameter. The bridge is oriented east-west. Each 40-m-long span is an arch strengthened with two struts. The arch and strut are both supported on the pile cap of drilled shafts. The bridge deck is integrally cast with the top of strut and arch but seated at the bent caps and abutments. Expansion joints exist at each abutment and bent.

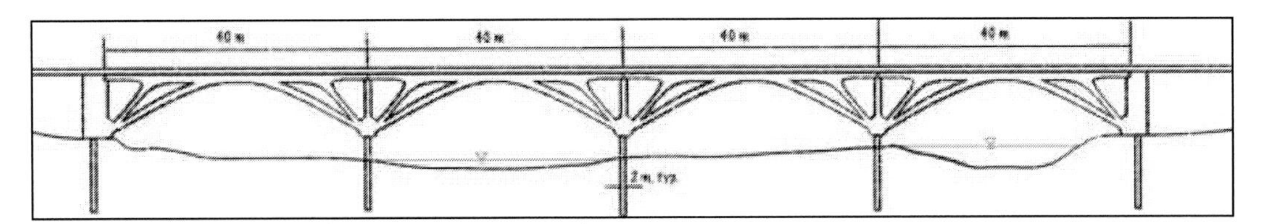

Fig 3.51 Elevation view of the Xiaoyudong Bridge

The two spans on the west side of the bridge had collapsed (Fig. 3.50), and the eastern most span was severely damaged (Fig. 3.49). The remaining span suffered little damage. The central bent is located in the middle of the river with the longest pile shafts exposed above the ground. This central bent was significantly tilted while the other intermediate bents appeared to have little rotation. The difference in the stiffness of the various bents most likely contributed to the collapse of the two western spans. Under severe shaking around the bridge, the central bent rotated, unseating the two spans. The bent cap was approximately 850 mm across; the seat length of each girder was approximately 300 mm, which is very limited for strong shaking at the bridge site.

The Caltrans reconnaissance team did not find any obvious evidence of lateral spreading and liquefaction at the bridge site. However, it is highly possible that under strong shaking, lateral spreading or liquefaction could occur in the middle of the river on the west side of the bridge. In that case, the tilting of the central bent can be easily explained.

The eastern most span also suffered significant damage. Shear failures were observed on the top end of the struts and on the bottom end of the arches. The shear key on the east abutment was damaged. At the top end, each strut seems to be reinforced with 15 #25 bars (Metric unit). Some were fractured.

The damage to the eastern most span was likely caused by the surface fault near the east abutment. Figure 3.52 presents a schematic view of fault rupture versus the location of the east abutment of the bridge. When the hanging wall of the thrust fault moved up right relative to the foot wall that supported the east abutment, the levee began to bear on the arches and added significant shear forces and bending moments especially at the east ends of the arches. Designed for axial forces, the arches sheared at their east ends. The top end of the arches remained intact due mainly to its larger section. Because of the shear failure in the arches, the east span deflected downward significantly and added more load to the strut, resulting in its shear failure as well. At the same time, Span 4 pushed toward Bent 4 and caused flexural cracks at the pile shaft.

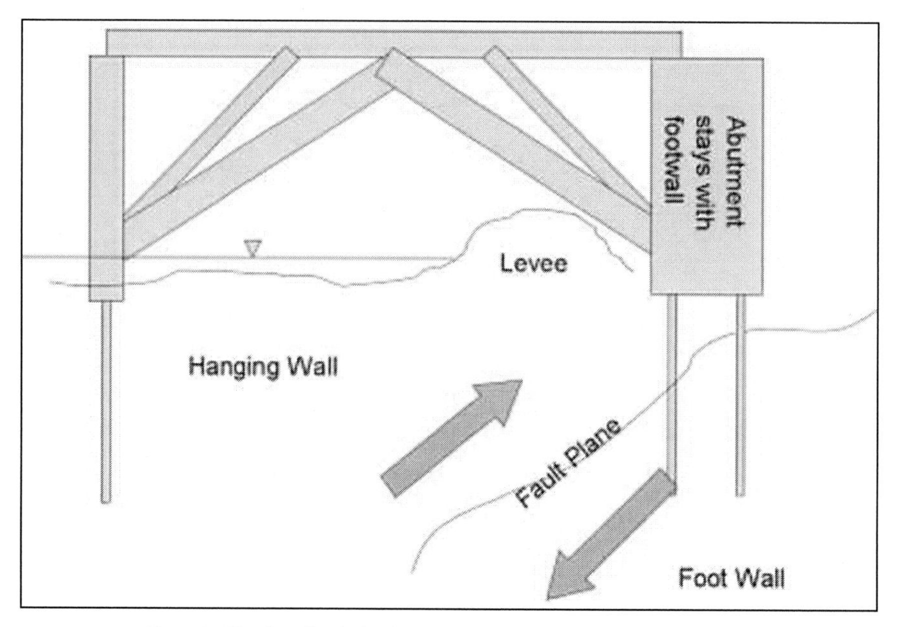

Fig. 3.52. Surface faulting and interaction with bridge

Spandrel Arch Bridge in Bailu of Pengzhou City (31.1807°N, 103.8859°E)

Figure 3.53 shows a collapsed stone-and-earthen arch bridge (also called the Sino-French Bridge). One span collapsed into the river (right side of figure), the other side remains standing. The commonly expected failure mode for unreinforced masonry arch bridges occurs when the seismic inertial loading is high enough to cause an opening in the arch's stonework (overcoming gravity compression); we would not be surprised if this failure mode occurred for this bridge.

3.9 Bridges in Beichuan County

Figure 3.54 shows a bridge crossing the Bai Cao He in Beichuan. All the spans were shifted to the right of the photograph and one span collapsed into the river. The lateral shear keys on the pier head were damaged; fortunately, the shift was insufficient to unseat the spans. Note the colocated telecommunication cables (most likely a fiber optic cable) were knocked off the end of the bent caps (Fig. 3.54). The tunnel on the opposite side of the river collapsed behind the portal (insert in Fig. 3.54).

Figure 3.55 shows a pedestrian suspension bridge over the Yu River in Baichuan County. The main span of the bridge was approximately 130 m long and 2.5 m wide. The 10 cm diameter suspension cables were supported by large, stiff, reinforced concrete towers. The deck suspenders were 2 cm in diameter. The superstructure consisted of two steel pipe girders and a wooden deck. Suspension bridges typically perform well during earthquakes. These long-period structures are strong and flexible and usually have little damage except to the deck, which can get banged around during an earthquake. During this earthquake, at least one of the towers moved toward the river either due to lateral spreading or from a surface offset (Fig. 3.55). The three enormous slides in Beichuan destroyed a significant portion of the building stock in the city as well as other bridges.

Fig. 3.53. Collapsed stone-and-earth bridge (Bailu town of Pengzhou City)

Fig. 3.54. Side view of damaged bridge near Beichuan County

Fig. 3.55. Beichuan Bridge after the Wenchuan Earthquake

Nanba Bridges (N32.2089°, E104.8277°)

An old bridge in Nanba collapsed (Fig. 3.56) along with a new bridge under construction just west of the old bridge (Fig. 3.57). A temporary bridge (Fig. 3.58) was constructed by the PLA to provide supplies to the communities in the mountainous region. A sign posted on the approach read "one car at a time"; however, it seemed that no one had the patience to wait (Fig. 3.59).

Surface faulting occurred in Hejiaba (Fig. 3.60), just east of Nanba across the river. An old bridge between Nanba and Hejiaba collapsed (Fig. 3.76). A temporary wooden bridge for foot traffic (Fig. 3.62) was constructed by the PLA, and military guards were posted to ensure no vehicle used the bridge. Along with traffic closure, the collapse also damaged a fiber optic cable on the bridge; a temporary cable was supported by a tripod made of wooden sticks (Fig. 3.63).

3.10 Arch Bridge in Zhongba

The arch bridge in Zhongba showed evidence that the ground shaking was sufficient to open cracks in the arch (Fig. 3.64 and 3.65).

Fig. 3.56. The old Nanba bridge collapsed leaving only part of its mid-pier standing in the middle of the river

Fig. 3.57. All spans of this partially completed nine-span bridge fell into the river except the two spans on abutments

Fig. 3.58. The temporary bridge built by the PLA was simply supported on both ends

Fig. 3.59. The temporary bridge was wide enough for single-lane traffic

Fig. 3.60. Surface fault in Hejiaba; the road used to be level and a few houses on the right of the photo collapsed

Fig. 3.61. View from Hejiaba towards Nanba of collapsed bridge and construction of new abutment

Fig. 3.62. A temporary wooden bridge was built for villagers to cross from Hejiaba to Nanba. Four PLA soldiers were posted on either end of the bridge to guard against vehicle crossing.

Fig. 3.63 Temporary fiber optic cables for telecommunication were installed to replaced the damaged cables due to bridge collapse (Fig 3.76)

Fig. 3.64. Arch bridge sustained damage, Zhongba Village of Pengzhou City

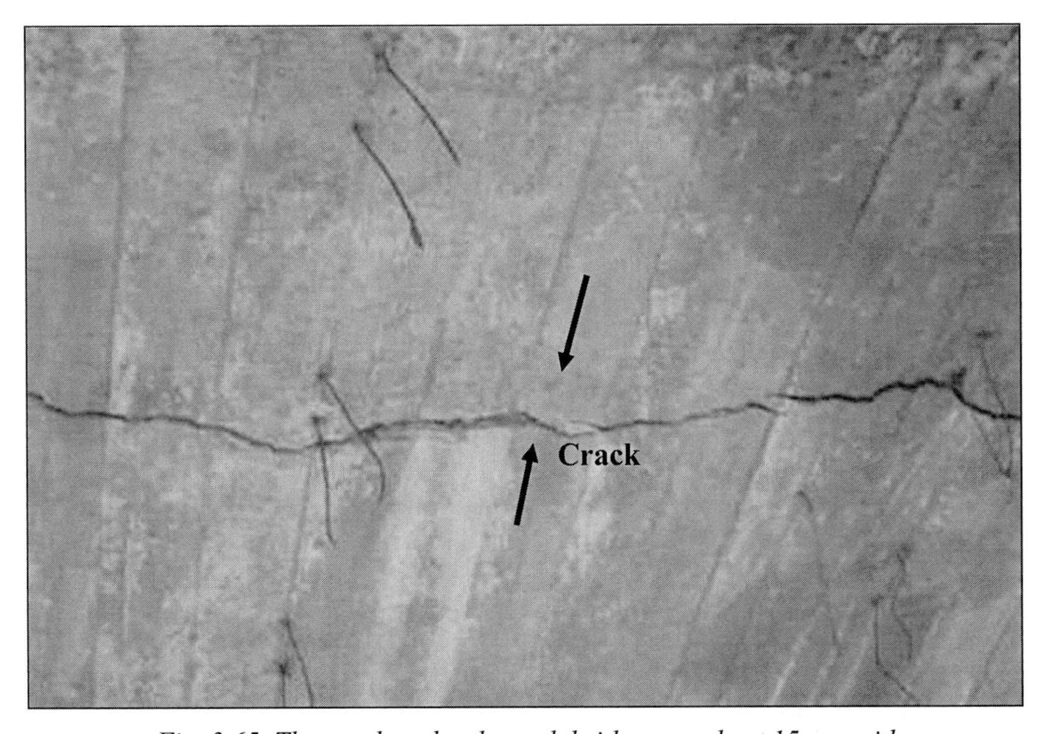

Fig. 3.65. The crack under the arch bridge was about 15 mm wide

3.11 Bridges of Dujiangyan

Dujiangyan is the closest major city to Chengdu. It was one of the large cities in the earthquake-affected area that sustained extensive damage to buildings, roads, power, telecommunication, and water. Ground motions ranged from a PGA of 0.20 g (southeast side of city) to 0.35 g or more in the northwest side of the city.

The city's key attraction is its flood management facility on the Min Jiang. There are two flexible bridges (Fig. 3.66 and 3.67) for pedestrians on this site. They were closed in July during the team's visit and were opened later after a safety inspection.

Damage to the stone railings demonstrated strong shaking in this area (Fig. 3.68).

Fig. 3.66. Flexible bridge in the flood management facility

Fig. 3.67. Flexible bridge in the flood management facility

Fig. 3.68. The poles of the stone railing close to the flexible pedestrian bridge were sheared off the stress riser close to the base

The unreinforced liners along the embankments of the river showed nearly continuous cracks, suggesting an average of 2 to 5 in. of settlement at the top of the embankments, coupled with 0- to 3-in.-wide cracks observed at the top of the embankments. The extent of cracking increases as the river gets closer to the rupture zone, presumably due to higher ground motions. The largest amount of cracking coincided with the most prevalent locations of broken stone railings.

3.12 Tunnels

During the field trip we saw a few tunnels and observed minor damage. During a October 2008 workshop in Chengdu, China Rail South West Research Institute (CRSWRI) provided us with information on a number of tunnels they had performed post-earthquake inspections on. We have since obtained additional information from ZhengZheng Wang at the Southwest Jiaotong University in Chengdu (Wang 2009). As shown Figure 3.69, almost all of the tunnel damage occurred south of Wenchuan.

Most of the damage was caused by landslides or by fault offsets, resulting in:

1. portal landslide blocking the entries and exits (most common);
2. collapsed and cracked tunnel liner (limited to zone of offset);
3. road surface upheaval; and
4. portal wall damage.

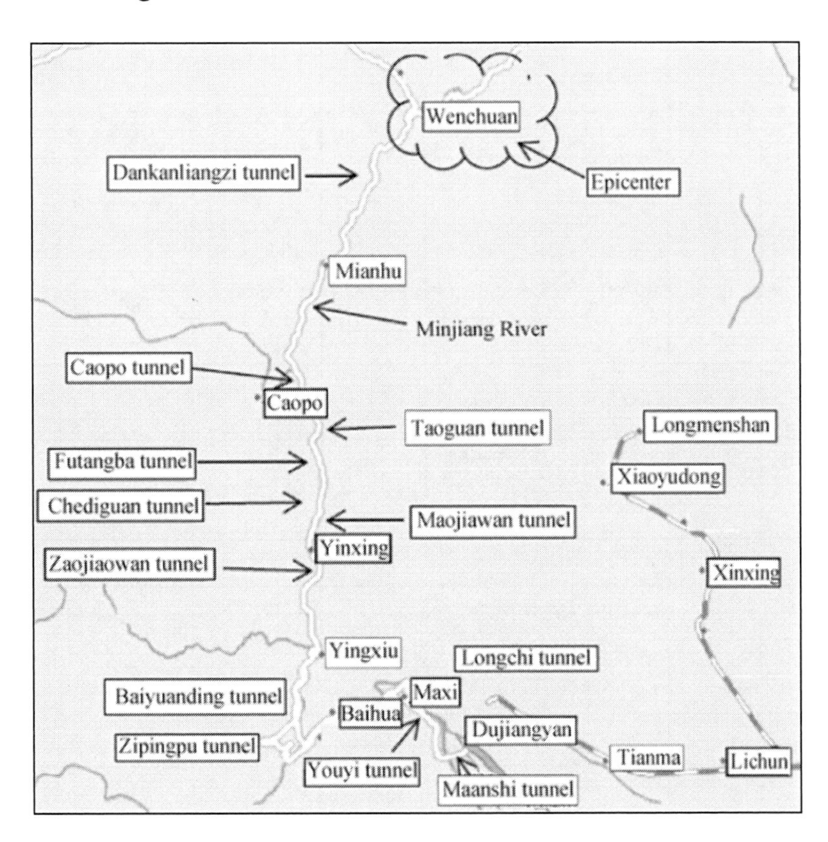

Fig. 3.69. Location map of tunnels investigated after the May 12, 2008 earthquake (drawing courtesy of ZhengZheng Wang)

Table 3-2. Tunnel Damage

No.	Tunnel name	Length (m)	Damage experience before earthquake	Tunneling method	Distance to the epicenter (km)	Distance to nearest fault (km)	Adverse geological	Damage
1	Caopo tunnel	759	——	Bench cut method	25	1 (WM Fault)	Fault	Portal failure and flooding
2	Maojiawan tunnel	399	——	Bench cut method	35	6 (WM Fault)	Fractured zone	Portal failure, rockfalls, and damage to liner
3	Longdongzi tunnel	1071	——	NATM	30	-	Fractured one	Portal failure, rockfalls, damage to liner, and flooding
4	Futang tunnel	300	——	Bench cut method	45	10 (WM Fault)	——	Portal failure and rockfalls
5	Taoguan tunnel	625	——	Bench cut method	28	6 (WM Fault)	Fault	Portal failure, rockfalls, and damage to liner
6	Jianmenguan kou tunnel	-	-	-	-	-	-	Fault offset caused damage to pavement and liner
7	Baiyunding tunnel	451	——	Bench cut method	50	0 (YB Fault)	——	Liner damage
8	Youyi tunnel	950	Gas explosion	NATM	52	0 (YB Fault)	Gas	Pavement cracks and liner damage
9	Longxi tunnel	3691	Cave-in and collapse	Full face method	49	-	Gas and fault	Portal failure, rockfalls, damage to liner, and pavement cracks
10	Shaohuoping tunnel	451	——	NATM	40	-	Fault	Portal failure, rockfalls, and damage to liner
11	Longchi tunnel	1177	——	Bench cut method	50	0 (DJ Fault)	Fault and gas	Portal failure, rockfalls, damage to liner, pavement cracks, and flooding
12	Zaojiaowan tunnel	1926	——	Bench cut method	42	12 (WM Fault)	Gas	Portal failure, rockfalls, damage to liner, and flooding
13	Zipingpu tunnel	4090	Gas explosion	Bench cut method	50	0 (YB Fault)	Gas and fault	Portal failure, rockfalls, and damage to liner
14	Chediguan tunnel	403	——	Bench cut method	32	12 (WM Fault)	-	Portal failure, rockfalls, damage to liner, and pavement cracks
15	Futangba tunnel	2365	Rock burst	Bench cut method	30	12 (WM Fault)	High ground stress	Portal failure, rockfalls, damage to liner, and flooding
16	Dankanliang zi tunnel	1567	——	Bench cut method	10	0 (WM Fault)	Fault	Portal failure
17	Maanshi tunnel	282	Collapse	Bench cut method	55	1 (DJ Fault)	Gas	Damage to concrete liner and flooding
18	Yingxiu tunnel	40	——	Bench cut method	48	-	——	Portal failure and rockfalls
19	Feishaguan tunnel	100	——	Bench cut method	50	-	——	Portal failure and rockfalls

Fig. 3.70. Portal landslide, Caopo Tunnel in Duwen (Courtesy CRSWRI)

Caopo Tunnel in Duwen

A landslide at the portal blocked the tunnel and the road leading to it (Fig. 3.70). The slope in front of the tunnel was not designed to prevent landslide as result of earthquake.

Maojiawan Tunnel in Duwen

The rock fall in front of the portal blocked the tunnel and covered the road leading to it. Accessing the tunnel was nearly impossible (Fig. 3.71); note that the reconnaissance team had to scale the slope to get to the tunnel. Although meshed wired net was used to prevent rock falls, it was not strong enough to keep the rocks from covering the road and the portal area.

Longdongzi Tunnel in Duwen

As shown in Figure 3.72, the left portal of the Longdongzi tunnel is almost entirely covered by a landslide, which fell from above the portal.

Fig. 3.71. Portal landslide, Maojiawan Tunnel in Duwen (Courtesy CRSWRI)

Fig. 3.72. Portal landslide, Longdongzi Tunnel in Duwen (Courtesy CRSWRI)

Futang Tunnel in Duwen

This portal area was also covered by the landslide. The first two spans were also covered by the landslide (Fig. 3.73).

Taoguan Tunnel in Duwen

The cracking of the portal wall was caused by the landslide above it (Fig. 3.74). Note that there were crib walls used to protect the road and the tunnel from landslide.

Fig. 3.73. Portal landslide, Futang Tunnel in Duwen (Courtesy CRSWRI)

Fig. 3.74. Cracking of Taoguan Tunnel in Duwen (Courtesy CRSWRI)

Jianmenguankou Tunnel

The concrete structural wall (right side of Fig. 3.75) provided protection against landslide. The damage on the left side seemed to be fascia that had fallen off, most likely because it lacked adequate anchoring capacity. Fault offset within tunnels caused some relatively modest damage. The exhaust fan was used to circulate air for the repair crews inside.

Baiyunding Tunnel

Figure 3.76 shows the liner damage close to the foot.

Fig. 3.75. Damage to portal of the Jianmenguankou Tunnel (left side of photo) (Courtesy CRSWRI)

Fig. 3.76. Damage to liner, Baiyunding Tunnel (Courtesy CRSWRI)

Youyi Tunnel

Figure 3.77 showed damage to tunnel liners.

Longxi Tunnel

The Longxi Tunnel runs along the Duwen highway, a cut-and-cover construction road. Figures 3.78 to 3.84 show cracking to the tunnel liners and upheaval of road surface.

Much more information about tunnel damage can be obtained from ZhengZheng Wang's excellent report (Wang 2009), which along with the other references in this chapter can be obtained from *www.springerlink.com*.

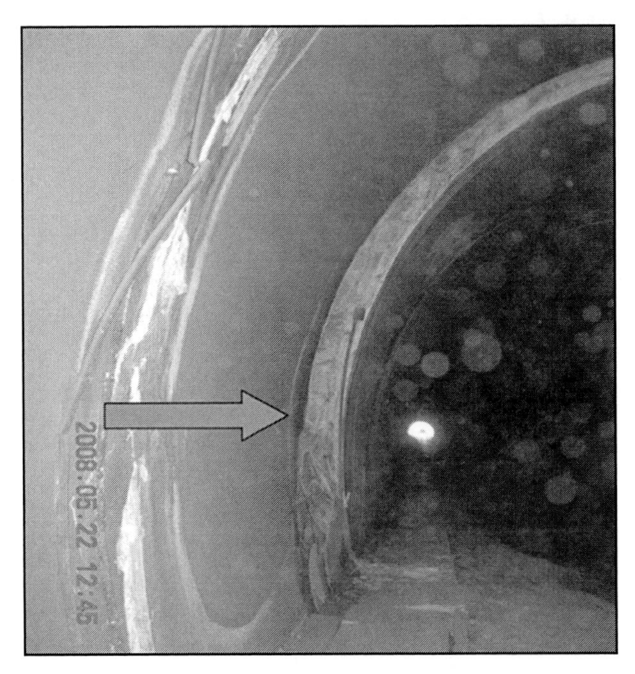

Fig. 3.77. Damage to Liner, Youyi Tunnel (Courtesy CRSWRI)

Fig. 3.78. Damage, Longxi Tunnel approach (Courtesy CRSWRI)

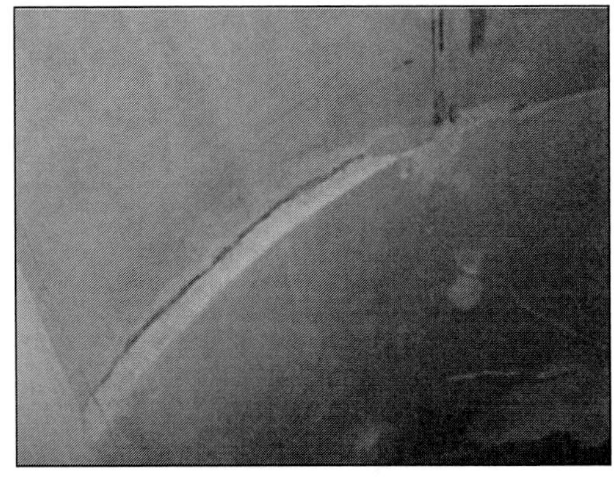

Fig. 3.79. Liner damage, Longxi Tunnel (Courtesy CRSWRI)

Fig. 3.80. Upheaval damage by 70 cm (Courtesy CRSWRI)

Fig. 3.81. Upheaval damage by 70 cm (Courtesy CRSWRI)

Fig. 3.82. Vertical movement damage to road surface in the tunnel (Courtesy CRSWRI)

Fig. 3.83. Repairing liner damage (about 15 cm) (Courtesy CRSWRI)

Fig. 3.84. Liner damage (Courtesy CRSWRI)

Fig. 3.85. Rockfall on railroad tracks (Courtesy of Prof. Mei and Dr. Zhang)

3.13 Rail System

The railway system also experienced extensive damage at four places in Baocheng, four in Chengkun, and seven in Chengyu. Additionally, a number of railway branches in Guanyue sustained various degrees of damage. A freight train from Baocheng to Yuguan derailed in tunnel 109 and was trapped. Another freight train in Xuixian County in Gansu Province derailed in a tunnel, and a freight car carrying flammable liquid caught fire in the tunnel. Rockfalls caused problems for railways just as they did for highways (Fig. 3.85). There was also damage to tracks due to liquefaction (Fig. 3.86), damage to railway bridges due to ground shaking (Fig. 3.87 and 3.88), and damage to railway tunnels due to landslides and fault offset (Fig. 3.89).

3.14 Recommendations

As China did not implement appropriate seismic design criteria for the area's hazardous environment, the May 2008 Wenchuan earthquake severely taxed the country's emergency response resources. The Longmen Shan Mountains have the steepest grades of any mountains in the world, and the May 12 Wenchuan earthquake only increased the region's vulnerability to rockfalls, landslides, debris dams, mudslides, and such.

Landslide damage to roads during earthquakes is considered inevitable in mountainous areas and apparently was not properly addressed by seismic criteria. Building mountain roads in tunnels and using more rock anchors could possibly reduce their vulnerability during large earthquakes. Protecting tunnels from landslides and fault offsets must be part of the design and construction criteria. In California, modern tunnels are designed and constructed for landslides and fault offset.

Fig. 3.86. Soil liquefaction damage to tracks (Courtesy of Prof. Mei and Dr. Zhang)

Fig. 3.87. Collapsed railroad bridge in Shifang (Courtesy of Prof. Mei and Dr. Zhang)

Fig. 3.88. Simple spans moved on their supports (Courtesy of Prof. Mei and Dr. Zhang)

Fig. 3.89. Gasoline tanker train derailed in a tunnel in Xuixian County in Gansu Province and caught fire (Courtesy of Prof. Mei and Dr. Zhang)

Obviously, the bridges exposed to strong ground shaking were designed for too low a ground motion, and we saw many examples of inadequate construction for earthquake loading. Bridges should be built where they are less vulnerable to landslides and fault movement where possible or designed to resist near-fault ground shaking. A multihazard approach to bridge design may prove useful. Where bridges must cross known active faults, suitable countermeasures should be taken. Designing bridges with increased strength (columns with 3 percent longitudinal reinforcement) and with ample spiral and hoop reinforcement should address strong ground motions. Designing the bridges to move without being pulled apart would address the fault offset hazard, but obviously, relocation is preferred.

We saw many damaged stone masonry closed spandrel arch bridges. These bridges need to be designed with great care in seismic areas. Unreinforced masonry arch bridges should not be used in high seismic areas. Arch bridges need to be made continuous between the abutments and should be designed for the demands of a much larger earthquake. More confinement reinforcement needs to be used in columns.

It would also be useful to increase the contact between bridge, railway, and highway engineers around the world who are trying to address similar issues. Retrofitting existing bridges appears practical. Restoring track to normal standards after an earthquake may require sizeable quantities of ballast. The vulnerability of track adjacent to coastlines (especially those on soils prone to lateral spreading or other types of geotechnical failure) should be considered in any earthquake response plan.

Given the existing infrastructure of bridges, tunnels, and roads in the area, it is doubtful whether China will find it cost effective to retrofit every one of these components of the transportation system in the Longman Shan Mountains or other mountain environments to be "earthquake proof." High-priority upgrade actions that are likely to be cost effective include retrofitting primary highway bridges to resist inertial loads due to PGA = 0.40 g without major collapse and reinforcing tunnel portals to limit the debris impact from landslides. More work could be done to map out landslide-prone zones along primary highways and, for the most at-risk locations, mitigate the slope to prevent the slide or limit its extent. This style of upgrade strategy should be implemented over a 10- to 20-year time frame for existing transportation links. Any new roads, tunnels, and bridges should be designed for at least PGA = 0.40 g, with suitable detailing for ductility and avoidance of slides as practical. New tunnels should not be installed through major active fault zones, and tunnel liners should be designed to accommodate up to 0.2 m of sympathetic faulting where potentially active faults are thought to exist. With this type of strategy, it should be possible to keep most roads, rails, tunnels, retaining walls, bridges, and other transportation facilities in service for smaller earthquakes and to limit the damage from very large earthquakes such that the critical links in the transportation network can be restored to service within a few days and within a month for all links.

3.15 Acknowledgment

The U.S. transportation structures team for the May 12, 2008, Wenchuan earthquake reconnaissance was coordinated and lead by Federal Highway Administration. The visits to various earthquake affected areas were made possible by the Ministry of Communication of China and the Sichuan Province Highway Planning, Survey, Design, and Research Institute. The support and help from the following persons and agencies are greatly appreciated:

1. The Federal Highway Administration
   Office of International Program: Mr. Ian Saunder, Director, and Mr. Steve Kern.
   Office of Infrastructure, R&D: Mr. Gary Henderson, Director; Mr. John McCraken; and Mr. Ian Friedland
2. The U.S. Embassy in Beijing

Financial support to complete the site visits and final report were provided in part by the American Society of Civil Engineers (ASCE), the Earthquake Engineering Research Institute (EERI), the Center for Transportation Infrastructure and Safety (CTIS), the Geo-Engineering Earthquake Reconnaissance (GEER), and the Mid-America Earthquake (MAE) Center. Local transportation was provided by the Sichuan Province Highway Planning, Survey, Design, and Research Institute.

Thanks are also due to Mr. Weilin Zhuang, chief engineer of the Sichuan Province Highway Planning, Survey, Design, and Research Institute, for his hospitality in welcoming the reconnaissance team at the frontline of post-earthquake reconstructions. Without his support, the U.S. team would not have been able to go into the secure areas for detailed inspections of several bridges that offered critical findings.

The authors also gratefully acknowledge extensive help provided by Mr. Wei Zhou, president of the Research Institute of Highway, Ministry of Communication of China (RIOH); Jinquan Zhang, vice president of RIOH, and Zhifeng Yang, director of the Science and Education Department of RIOH; and Xuelei Zhu, director of the Science and Education Department of the Communications Department of Sichuan Province.

The authors appreciate the information on tunnel damage provided by China Rail South West Training Institute and by Mr. ZhengZheng Wang of Jiatong University. A better understanding of landslides, debris slides, and avalanches (along with the use of maps, photos, and drawings) from Mr. Zhaoyin Wang at Tsinghua University and Mr. Yueping Yin form the China Geological Survey was also much appreciated.

3.16 References

Deren Li, "Remote sensing in the Wenchuan earthquake." Photogrammetric Engineering & Remote Sensing, May 2009, p. 506-509.. Liu Ziqiang and Shuqin Sun. "The disaster of May 12 Wenchuan Earthquake and its influence on Debris flows." Chengdu University Journal of Geography and Geology, Vol. 1. No. 1, May 2009

Tang, C., et. Al., "Rainfall-triggered debris flows following the Wenchuan earthquake," Bulletin of Engineering Geology and the Environment. 2009

Wang, Zhaoyin, "Avalanches, landslides, and quake lakes induced by the Wenchuan earthquake on May 12, 2008.' Tsingha University, 2009

Wang, ZhengZheng, et. al, "Investigation and assessment on mountain tunnels and geotechnical damage after the Wenchuan earthquake." Science in China Series E, vol. 52. no. 2, February 2009

Zhang, Zuxun, et, al., "Photogrammetry for first response in Wenchuan Earthquake," Photogrammetric Engineering and Remote Sensing, May 2009

4 ELECTRIC POWER SYSTEM

OVERVIEW

The Sichuan Province State Grid is the primary operator of the high-voltage transmission and low-voltage distribution electric systems serving more than 46 million people in Sichuan province. Power generation includes thermal (nuclear, coal) and hydroelectric power, with the owners of the generation facilities including other companies and agencies.

The damage to the State Grid electric system in Sichuan province is listed in Table 4.1. The information reflects the damage and repair efforts through October 22 2008. The column "Repaired as of 10/22/2008" reflects the replacement and repair of equipment at existing substations/circuits; most of these repairs covered replacement of damaged circuit breakers, replacement of bushings, and such. The column "Repair in place beyond 10/22/08" reflects relatively modest repair efforts remaining where local conditions have prevented completion. The column "Reconstruct entirely beyond 10/22/08" indicates substations with nearly complete damage (such as the collapsed dead-end towers at Ertaishan shown later in this chapter), those that were impacted by landslide, or those that serve communities that are essentially destroyed/depopulated. These are considered beyond repair by the State Grid and will need outright replacement. The rows "10 kV Circuits and Substations" primarily indicate where landslides or surface faulting pulled down major lengths of low-voltage distribution, with pole mounted transformers called substations.

Table 4.1. Damage to State Grid Electric System, Sichuan Province

Voltage (kV)	Type	Total	Damaged	Repaired as of 10/22/2008	Repair in Place (beyond 10/22/08)	Reconstruct Entirely (beyond 10/22/08)
500	Substation	18	1	0	0	1
500	Transmission Line	41	4	2	2	2
220	Substation	94	13	9	1	4
220	Transmission Line	337	46	24	3	19
110	Substation	351	66	61	0	5
110	Transmission Line	796	118	117	0	1
35	Substation	351	91	84	0	7
35	Transmission Line	603	106	106	0	0
10	Substation	5473	795	749	0	46
10	Circuits	5876	1700	1606	0	94

Source: Sichuan Electric Power

Table 4.1 does not include damage to privately owned or non-state-grid substations, such as those high electric power demand manufacturing facilities (the aluminum manufacturers shown in Fig. 4.24 through 4.29, for example). Additionally, Table 4.1 does not reflect the geographic distribution of the substations. For example, no 500 kV substations were located within 15 km of the primary co-seismic fault ruptures. Instead, most experienced ground shaking under PGA = 0.10 g, so the relatively low damage rate for 500 kV substations should not be interpreted as the 500 kV equipment being well designed for seismic loading. Based on field observations by the ASCE/TCLEE team, 100 percent of 110 kV and 220 kV substations that experienced ground motions exceeding PGA = 0.3 g experienced functional damage to at least 15 percent of the equipment at the substation (sometimes nearly 100 percent).

Of the 171 substations (35 kV to 500 kV) that had at least some damage, 17 substations were "completely damaged" and are listed to be entirely rebuilt beyond October 22, 2008. Among the "completely damaged" substations we observed, there were many dead-end tower collapses, which essentially damaged all yard equipment below. Apparently, one substation was impacted by a landslide; some control and other buildings had collapsed, including a groundskeeper building. Of these 17 substations, five are planned to rebuild at the same location (two 220 kV, two 110 kV, and one 35 kV), while 12 will be rebuilt at other locations (one 500 kV, two 220 kV, three 110 kV, and six 35 kV).

The Sichuan Province State Grid has estimated that the cost for repairs and replacements to the electric system will be 31.3 billion RMB (US$4.6 billion). It also estimates an additional economic impact of 10.65 billion RMB (US$1.56 billion); this likely reflects a loss of revenue due to the inability to sell power as a result of the damage, plus the reduction in regional economic activity due to damage to industrial, commercial, and residential customers.

Table 4.2 summarizes the damage to equipment owned by Sichuan Electric Power Company (excluding equipment owned by others). In Table 4.2, the "Number Damaged" column would show "3" if three single-phase transformers had been damaged; or "1" if one of the three-phased transformers had been damaged. The voltage level is provided where know; for circuit breakers, most of the damage would be in 220 kV (65 percent) with the remaining damage in 110 kV (35 percent) approximately.

Of the 116 damaged power transformers, 76 had damaged radiator pipe connections resulting in oil spills, and 33 slid away from their original positions. Other damage observed to power transformers by the ASCE/TCLEE team including failed bushings and pulled underground cables. There may have also been damage to lightning arrestors, oil conservators, and internal cores, but this remains to be confirmed by future detailed studies.

The power generation plant in Yingxiu was damaged and was not back in service on July 16, 2008. In some towns visited, there was up to 10 days of electric power outage. West of Beichuan, power was restored in 10 days (PGA ~ 0.25 g). There was substantial damage observed to low-voltage distribution systems (by landslide, by faulting, and some by inertial ground motion) as well as to high-voltage substations (by inertial ground motion) (see Table 4.1).

Table 4.2. Damage to Electric System Components

Voltage (kV)	Type	Number Damaged
500	Power Transformers (PT)	7[1]
220	Power Transformers (PT)	25
110	Power Transformers (PT)	84[2]
110	Current Transformers (CT)	115
220	Potential Transformers (PT)	16
500	Potential Transformer (PT)	1
--	Circuit Breakers (CB)	91
--	Disconnect Switches (DS)	Many

By June 10, 2008 (29 days after earthquake), 155 of 171 35 kV or higher power substations were repaired and restored to service, and 2,607 of 2,769 transmission and distribution lines (10 kV to 500 kV) were restored to service. Of the 17 completely damaged substations, the 220 kV substation in Anxian, 220 kV substation in Dakang, 110 kV substation in Xiaoba, 100 kV substation in Yuanmenba, and 35 kV substation in Jujiaya were planned to be rebuilt by August 31, 2008. As of October 18, 2008, two substations in Yingxiu (220 kV and 110 kV) remained in their immediate post-earthquake damaged condition.

Figures 4.1 and 4.2 provide satellite night-time coverage of the areas with light. The major city of Sichuan, Chengdu, in the brightest zone toward the bottom center of the images had electric power. The many other zones nearby had power outages.

4.1 Electric Power Generation

In 2005, the overall electric power generation distributions in China were:

- 76 percent fossil fuel (coal);
- 7 percent hydro (dams);
- 3 percent nuclear; and
- 14 percent wind, solar, geothermal, and such.

- A hydroelectric power plant near Nanba was out of service. Figure 4.3 shows the damage (large diagonal cracks) in exterior structural walls (also note the collapsed fence in the foreground). Permanent ground deformations of perhaps a few inches of settlement were prevalent around the building. Figure 4.4 shows the substation at the power plant; note the circular concrete-style dead-end towers over a step-up

[1] Includes one 500 kV Reactor

[2] Includes one 110 kV Reactor

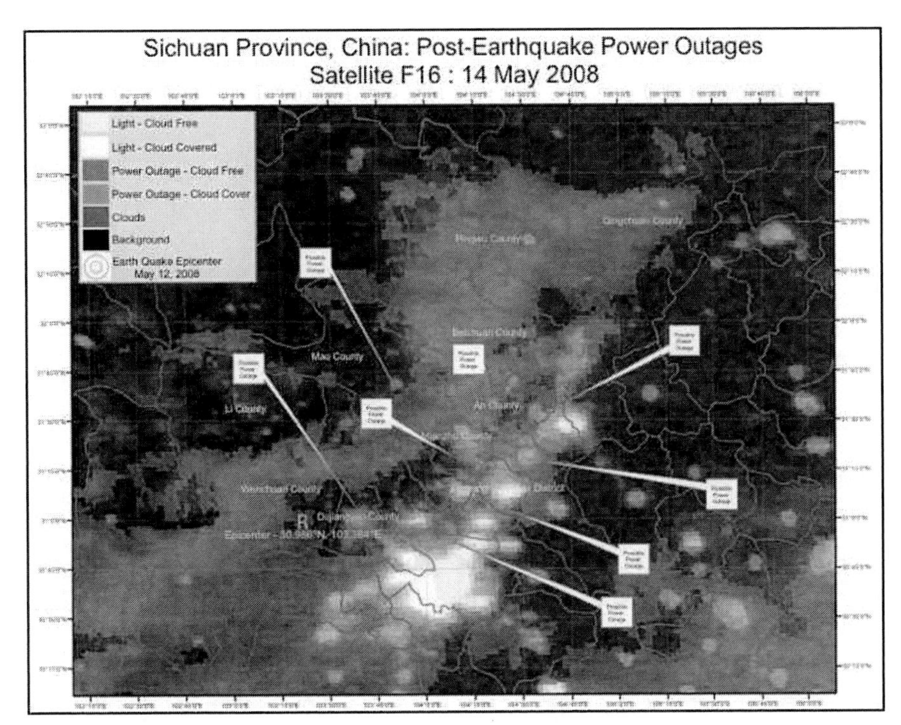

Fig. 4.1. Satellite map of areas with light, May 14, 2008

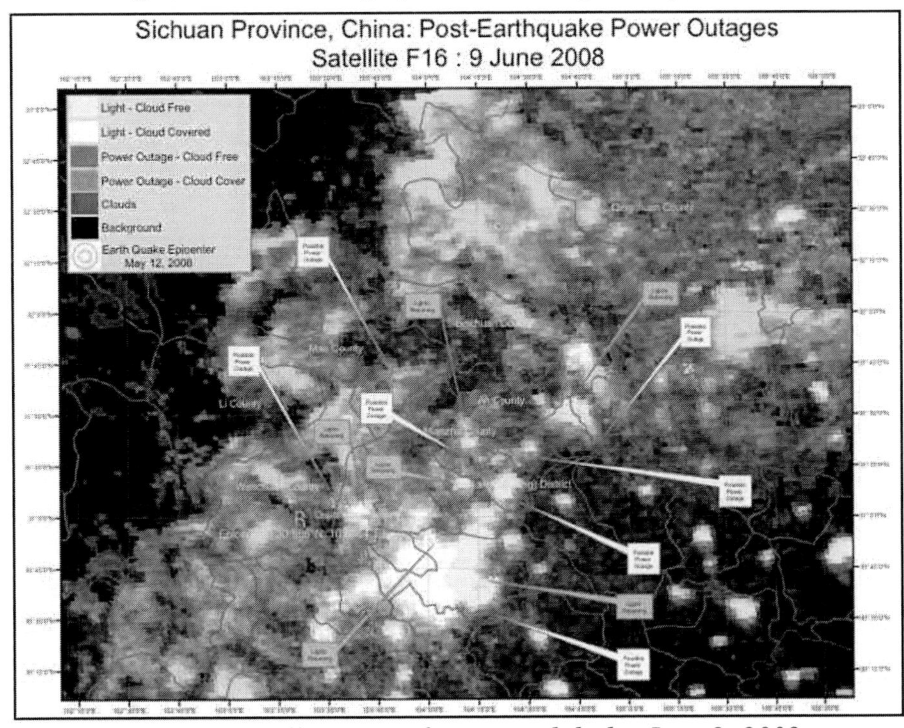

Fig. 4.2. Satellite map of areas with light, June 9, 2008

Fig. 4.3. Hydroelectric power plant building (Nanba)

Fig. 4.4. Substation at hydroelectric power plant building (Nanba)

transformer. The high-voltage bushings on the transformer appeared new, but we were not able to enter the site for closer inspection. At 220 kV substations that experienced higher levels of ground shaking (PGA > 0.4 g), these types of dead-end towers collapsed, resulting in gross damage to equipment below. Within 150 m of the power plant was a newly constructed 115 kV transmission tower, suggesting that a landslide had failed an earlier tower near the plant.

The ASCE team drove past the plant in Jiangyou, but did not attempt to enter. It was reported that the plant was not affected by the earthquake and was observed to be in operation on July 17, 2008. We do not know what seismic level the plant was designed for.

Figure 4.5 shows the powerhouse, substation and control building at the base of the Zipingpu Dam. ASCE team investigated these buildings. We estimated that at the base of the dam, ground motions were about PGA = 0.2g. There are four three-phase 500 kV transformers, all unanchored; one slid several inches and broke its 500 kV side riser (gas insulated). The 500 kV gas-insulated indoor substation performed well. One 500 kV lightning arrestor and one 500 kV CVT (Current Voltage Transformer) on the roof of the substation building broke. The building structures at the base of the dam performed well, with minor structural damage. The four 190 MW turbines (total 760 MW) all had various forms of damage and each had to be overhauled in the year after the earthquake.

Atop the 162-meter tall rock-fill dam, there was much more serious damage. The upstream face of the dam suffered damage to water seals. At the crest, the dam settled a little under 1 meter, within a Chinese assumed allowable of 1% of dam height. The settlement at the crest grossly failed the upstream water parapet wall, which had to be replaced in its entirety.

The radial gate for the dam's only spillway was inoperable after the earthquake, owing to distortions of the dam. As the reservoir was not at flood stage, this in itself did not cause any immediate trouble.

The crane over the outlet structure with four tunnels to four turbines and a fifth sedimentation tunnel was damaged. This crane is located adjacent to the crest of the dam. The motors to roll the crane broke off. Fortunately, at the time of the earthquake, the crane was "hooked" onto a closed slide gate below, and the combined crane/gate configuration prevented the crane from jumping off its rails. However, the net result was that the crane could not be used to open the gate to the sedimentation tunnel. Water could go to the turbines, but at low rates, not enough to keep the reservoir from rising.

There are two bypass tunnels, each with its own control building and slide gate. The control buildings are located at the crest of the dam, where ground motions were likely well in excess of PGA = 0.5g (possibly more). The winch mechanisms for both slide gates were damaged, likely to due to differential settlements. Both reinforced concrete frame control buildings were heavily damaged, with gross cracking of infill walls.

Secondary faulting on a fault that goes through one of the outlet tunnels created damage in the tunnel (the tunnel liner failed, but the tunnel still had a flow path, albeit with high surface roughness which severely impacts its flow rate).

Immediately after the earthquake, with the damage to cranes and winch mechanisms, as well as the spillway radial gate, essentially meant that no water could go through the dam (except via the turbines). This meant that at the inflow rate of the upstream river, the dam had about 10 days of storage before it would be overtopped. Landslides to access roads were all but shut-off access to major equipment to make repairs. Fortunately, two employees who were stationed at the dam were able, within 27 hours, to repair one of the slide gate winch mechanisms, and open one of the slide gates to one of the bypass tunnels. By opening this gate, the outflow through the dam was nearly the same as into inflow, thus avoiding gross failure.

Since the 2008 earthquake, the dam operator has studied the impact of the number of upstream landslides into the river. Essentially, the landslide debris has filled in the river upstream of the dam, but still within the "dead water" zone. So, the debris has not impacted the ability to produce power, although storage and flood control capability has been reduced.

Interestingly, the failure of one of the upstream highway bridges (Figure 3.44) forced a multi-year detour while the bridge was repaired. As the detour road was at lower elevation, the dam operator was unable to fill the dam to its full operating level for 2 years; this led to lower hydroelectric output over this time frame. This shows that a lifeline interaction occurred here; whereas the insufficient seismic design of the highway bridge led to collapse; and this collapse then led to loss of hydroelectric power.

Since this earthquake, the seismic requirements for this dam have been "upgraded" from PGA = 0.1g to PGA = 0.2g. When the ASCE team visited the dam in August 2012, all buildings and cranes had been repaired, but we still observed unanchored 500 kV transformers. Given the proximity of this dam to major active faults (within 5 km or so), and secondary faults (through an outlet tunnel), it would appear that this dam would not pass safety-requirements in California, especially with regards to the ability to drain the reservoir after a maximum credible earthquake; however, the Chinese operator of the dam reported to the ASCE team that the dam, as of 2012, meets the most modern Chinese safety standards. The differences between US and Chinese safety standards for dams deserves further examination.

When the ASCE team visited in August 2012, Chinese dam engineers reported that there was a variety of damage to upstream "run-of-the-river" dams. This included overtopping of some dams (due to landslide impingements); boulders that rolled down hills and destroyed control buildings and emergency generators; fault offset through tunnels that closed off power-generation capability, etc. They reported that these dams still performed "well". Further study is required to understand their criteria of performing well based on the reported damages.

4.2 Transmission System

There was widespread damage to high-voltage substations due to strong ground motion as well as damage to transmission towers due to landslides.

Figure 4.6 shows some typical damage to a live tank circuit breaker. In this case, the initial damage is concentrated at the base of the porcelain, where it is connected to a flange with only a shallow embedment. In Figure 4.7, the porcelain failure is at the top of the bottom porcelain unit. The primary weaknesses of this type of circuit breaker are either excessive bending moments on the porcelain or weak porcelain-to-flange fittings.

Between Deyang and Shifang, the ASCE team observed a 110 kV substation (Fig. 4.8 and 4.9). We did not enter to inspect the equipment in detail as there was no attendant available, but we observed no damage from the gate. The nearby town of Deyang also had very minor damage.

Figure 4.10 shows a transformer on wheels that jumped its tracks. This type of damage has been observed in almost every corner of the world when ground motions exceed PGA = 0.25 g or so. This type of weakness is relatively easy to mitigate, and we would recommend that every wheel-mounted transformer in China where the 475-year earthquake PGA level is estimated to be 0.15 g or larger should be properly anchored. This is a prudent upgrade priority.

Figure 4.11 shows a slid 110 kV transformer at Chuanxindian in Shifang. The sliding exceeds 1 ft. for a steel-on-concrete installation. The staining on the concrete reflects that there was a large oil leak here (see Chuanxindian 110 kV Substation and Fig. 4.52 through 4.54).

Fig. 4.5. Hydroelectric power plant buildings (Zipingpu Dam) (Courtesy of IEM)

Fig. 4.6. Close-up of damaged live tank circuit breaker (110 kV)

Fig. 4.7. Damaged live tank circuit breaker (220 kV)

Fig. 4.8. 110 kV substation located between Deyang and Shifang

Figure 4.12 shows the sliding of a 110 kV transformer at Beichuan substation (PGA > 0.3 g). It leaked oil, most likely due to radiator pipe damage. While the two radiators are tied to each other, there appears to be no lateral support for the radiator other then the pipe connections. The State Grid reported many oil leaks due to pipe failures of radiators.

Fig. 4.9. 110 kV substation located between Deyang and Shifang

Fig. 4.10. Wheel-mounted transformer

Fig. 4.11. Skid-mounted transformer

Fig. 4.12. Skid-mounted transformer

Fig. 4.13. Collapsed control building

Figure 4.13 shows a collapsed control building at Anxian. The transformer on the right also slid, damaging grounds and likely imparting pull-down loads on adjacent disconnect switches.

Figures 4.14 and 4.15 show the damage to two bushings on a 500 kV transformer at Maoxian. The Phase C bushing caught fire. The Phase B bushing had total failure of its porcelain. These appear to be grouted bushings, which are rarely used in the United States. The adjacent surge arrestors had good slack and appeared unharmed.

Yingxiu 110 kV Substation

The Yingxiu substation was located in the western section of Yingxiu. Damage to nearby structures would suggest a PGA of about 0.5 g at the substation.

Figure 4.16 shows a schematic of the substation. We were not allowed to enter the substation, so all observations were made from outside the boundary fence. We estimate that we were able to observe about 80 percent of all equipment damage. The yard equipment damage is shown as dark gray dots (completely destroyed, requires replacement) or light gray dots (minor repair work to re-install the bus). The State Grid transmission company had not begun repair work as of 160 days after the earthquake, and there were weeds growing in many areas reflecting the lack of maintenance at the yard for more than 5 months. Most likely reason for the lack of repair is that the factories within a mile or so of the yard were heavily damaged, with no repair efforts ongoing, resulting in loss of load.

Fig. 4.14. Bushing, 500 kV transformer (Phase C burned)

Fig. 4.15. Bushing, 500 kV transformer (Phase B lost porcelain)

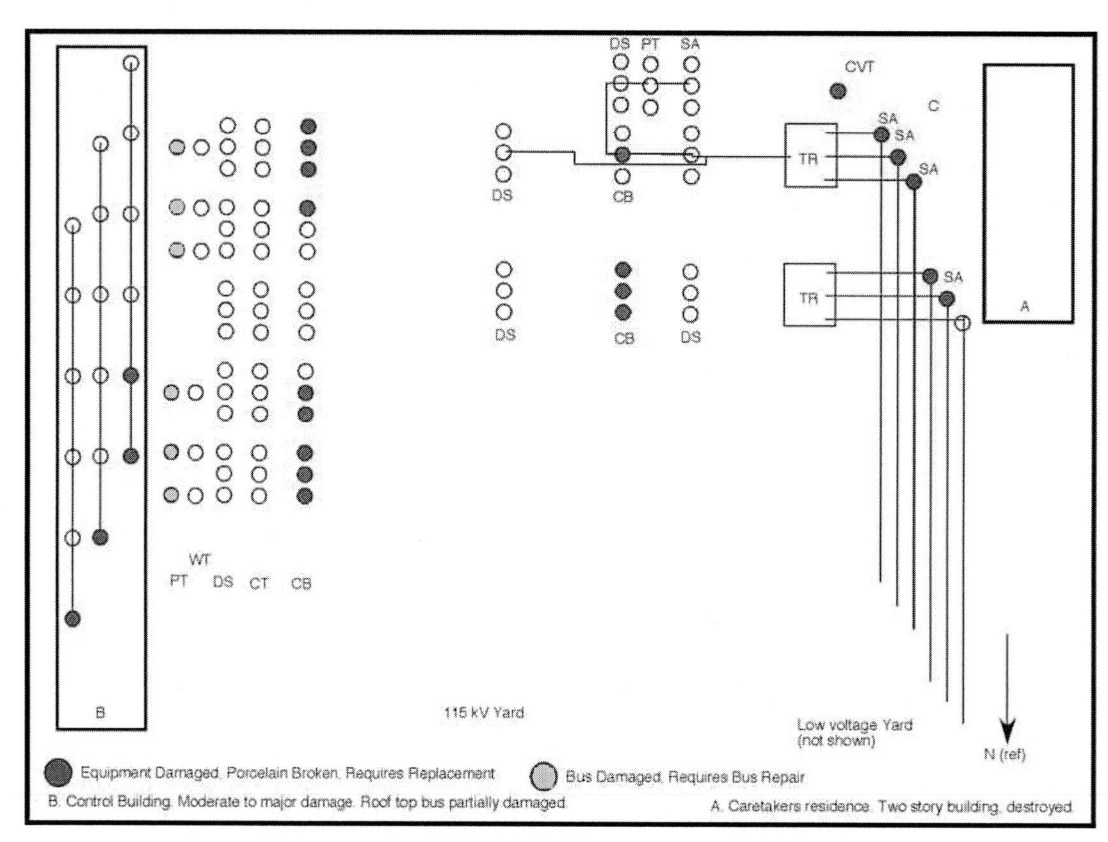

Fig. 4.16. Diagram of Yingxiu 110 kV substation

Fig. 4.17. Examples of damage to yard equipment, Yingxiu 110 kV substation

Fig. 4.18. Collapse of caretaker's two story URM residence

Fig. 4.19. Damage to current voltage transformer (CVT) and three lighting arrestors

We observed gross failure of at least 13 of 21 live tank circuit breakers. It is possible that all were damaged, but we could not gain close enough access to verify each individual breaker. This type of breaker was used at many substations (see also close ups in Fig. 4.6 and 4.7). The commemorative mural at the Beichuan 220 kV substation (Fig. 4.36) highlights this particularly weak breaker.

Figure 4.17 shows a different view of the damaged circuit breakers (CB) than Figure 4.16. The inertial failure of the live tank CBs grossly damaged the adjacent disconnect switch (DS) about 15 percent of the time but never visually damaged the adjacent current transformer (CT). The CT has a larger post than the DS, so it is relatively more strength than the DS to support the weight of the failed and collapsed live tank circuit breaker (LTCB).

There were six hanging wave traps; none visually appeared to be damaged. Five wave traps (WT) are shown in Figure 4.22. In Figure 4.22, all three bus drops to the three power transformers (PT) were broken, probably due to large movements of the hanging WTs and insufficient slacks that impacted the drop. Though less likely, it may also have been caused by cable dynamics.

Yingxiu 220 kV Substation at Aluminum Factory

The team visited an aluminum manufacturer in Yingxiu 160 days after the earthquake. This high-value industrial facility uses the electrolytic process. Private repair crews were at work at the time of the visit.

Fig. 4.20. Damage to live tank CBs; pull down of some adjacent DS

Fig. 4.21. Damage to live tank CBs

Fig. 4.22. Hanging wave traps and damage to bus connections to PTs. 15 percent of DS adjacent to hanging WTs were damaged due to pulldown

Fig. 4.23. Control building with large cracks on top right corner, small cracks in walls, and columns between windows

Figure 4.24 shows the anodes used in the electrolytic method. Presumably, these had been removed from their acid baths so that the tanks could be refurbished.

The production of aluminum ore from bauxite in this manner requires a large amount of electric current. This facility had five three-phase 220 kV transformers and two three-phase 110 kV transformers. When we visited the site on October 17 2008, two 220 kV transformers had been repaired, and five more 220 kV transformers were still in various stages of repair. We were not allowed to enter the facility and had only a short time to inspect the repairs being made to the 220 kV transformers.

We observed that 15 220 kV and 15 110 kV bushings and all radiators had been removed. At least three of the 10 oil conservators had been removed as well (Fig. 4.25). From the "boneyard" (Fig. 4.28, and 4.29), at least one 220 kV bushing was damaged beyond salvage (gross broken porcelain), and the porcelain of one other 220 kV bushing had apparent slippage from its metal support structure (Fig. 4.29).

We surmise that the oil conservator tanks had been removed to expedite replacement of the 220 kV and 110 kV bushings. We observed at least six shipping crates for new 110 kV bushings and six more shipping crates for new 220 kV bushings. It was evident that many of the original bushings were intended to be replaced.

The bushing damage was probably triggered by either sliding of the transformers or inertial loading, or a combination of both. All transformers we observed showed evidence of new anchorage to heavy concrete foundations.

Ertaishan Yingxiu 220 kV Substation

Figure 4.30 shows an aerial view of the destruction of Ertaishan substation in Yingxiu. The primary cause of the damage is inertial induced destruction of concrete dead-end structures, resulting in collateral damage to transformers and circuit breakers.

Fig. 4.24. Anodes for aluminum electrolysis

Fig. 4.25. Oil conservator tanks

Fig. 4.26. Radiators

Fig. 4.27. Transformer with new anchors; new 220 kV bushings in shipping crates

Fig. 4.28. "Boneyard" with damaged 220 kV bushings

Fig. 4.29. Slipped porcelain bushing is fourth from top

Fig. 4.30. Destruction to Ertaishan Substation, 220 kV, Yingxiu

Fig. 4.31. Destruction to dead-end towers, Ertaishan Substation

Fig. 4.32. Destruction to yard equipment, Ertaishan Substation

Fig. 4.33. Destruction to yard equipment, Ertaishan Substation

Fig. 4.34. Damage to Beichuan substation, 110 kV (May 2009 photo)

Beichuan 110 kV and 220 kV Substations

This substation was located in Beichuan, a small city with a pre-earthquake population exceeding 50,000, was heavily damaged by strong ground shaking (likely well over PGA of 0.4 g), and portions of the town were impacted by landslides and debris flow.

When the ASCE/TCLEE team visited the substation in May 2009 (one year after the earthquake), it had not yet been repaired or rebuilt as Beichuan had been designated by the government as a memorial site for the earthquake.

Figure 4.34 shows damaged transformer radiators (left) and a collapsed dead-end tower (pole lying down to right).

Figure 4.35 shows a collapsed current transformer that appears to have been impacted by the collapse of the top beam and masts of a dead-end structure (Fig. 4.36). Dead-end structures collapsed at several substations: one at this substation, one or more at the Beichuan 220 kV substation, three at Ertaishan 220 kV substation, and one at Yinxing substation (likely impacted by landslide). The cause(s) of these collapses remains speculative and might have been resulted from any of the following:

- inertial overload on the dead-end structure;
- pull-down forces caused by falling substation equipment with attached bus to the dead-end structure;
- collapse of other nearby items (possibly light poles and such) that might have impacted the dead-end structure; and
- impact from landslide.

Normally in the United States, these dead-end structures have been assumed to be seismically-robust in that they were elastically designed for unbalanced line loads plus wind and ice loads, which in almost every instance exceed the loads from even the largest earthquakes. Certainly, these collapsed dead-end structures in China should be further investigated to determine the cause of failure, as their impacts on nearby substation equipment resulted in substantial damage.

Figure 4.37 shows a mural outside a large 220 kV substation about 20 km west of Beichuan. PGA in this area was likely about 0.25 g. We were not allowed to enter this substation; however, the site superintendent suggested that overhead dead-end towers collapsed leading to major damage to the yard equipment. The mural highlights the various kinds of repair efforts. The hammer-and-sickle in the mural denotes the Chinese labor force.

Fig. 4.35. Damage to Beichuan substation, 110 kV (May, 2009 photo)

Fig. 4.36. Damage to Beichuan substation, 110 kV (May 2009 photo)

Fig. 4.37. Memorial erected at Beichuan Substation, 220 kV (Note the failed LTCB on the right side of John Eidinger, typical of all observed failures).

Mianzhu 220 kV Substation

The Mianzhu substation is located about 20 km southeast of the rupture, likely experiencing ground motions of PGA = 0.2 g or a bit more. The URM fence around the substation collapsed in one location (perhaps 5 percent of total fence) but did not damaging any equipment due to its impact. Several CVTs were damaged, likely due to swinging wave traps above them. It took 5 days to make repairs.

Caopo Power Station

Figure 4.38 shows a damaged live tank circuit breaker at the Caopo power plant. The current transformer to the left of the broken breaker has a taught cable holding up the broken breaker, and the discoloration of the CT shows oil leakage, likely due to the high load. On the other side, the breaker is attached to the remnants of a broken disconnect switch, likely broken due to pull-down.

Figure 4.39 shows damage to CVTs at the Caopo power plant.

Fig. 4.38. Circuit breaker with damaged adjacent current transformer and disconnect switches due to pulldown

Fig. 4.39. CVT damage

Figure 4.40 shows damage to the 110 kV live tank circuit breakers. These types of breakers have been damaged in past earthquakes around the world. They were prevalent in China and were damaged at many different substations. The damage to the horizontal break disconnect switches to the left of the breakers is most likely caused by pulldown of the failed breakers. The influence of the tight (little to no slack) bus in the failure of the breaker is uncertain, but once the breaker fails, the pulldown of the adjacent switch is not uncommon. Figure 4.40 shows a close up of the porcelain column for one of the damaged breakers. Notice the failed URM brick wall in the background, from which we assume that the local PGA must have been at least 0.30 g. The background of Figure 4.40 shows x-cracks in non-ductile concrete frame columns, further illustrating that the ground motion at this substation was likely in excess of PGA = 0.30 g.

Yuanmenba Substation

Figure 4.41 shows broken live tank circuit breakers at the Yuanmenba substation. These types of breakers had a very high failure rate (likely exceeding 75 percent) whenever exposed to a PGA > 0.35 g or so.

Collapse of control buildings led to the damage of low-voltage switchgear equipment within. Figure 4.42 shows the partially collapsed URM control house at Yuanmenba substation.

Fig. 4.40. Damage to live tank circuit breakers, Caopo Power Plant

Fig. 4.41. Live tank breaker damage, Yuanmenba Substation

Fig. 4.42. Control Building Equipment Damage, Yuanmenba Substation

Fig. 4.43. 220 kV live tank circuit breaker damage, Yongxing Substation, Mianyang

Yongxing Substation

Figure 4.43 shows damage to 220 kV circuit breakers. Note the extra height of the porcelain required for higher voltage installations as compared to the 110 kV installations. The taller height leads to higher overturning moments, and the larger diameter porcelain is still insufficient to resist the applied loads. The level of ground shaking at this location was likely under PGA = 0.15 g.

Jiangyou Power Plant

Figure 4.44 shows damage to ABB candlestick-type 220 kV circuit breakers.

Yinxing 220 kV Substation

The substation in Yinxing is located at the foot of a steep slope near the town, about 20 km northwest of Dujiangyan. Most of the damage to its equipment and structures was caused by a landslide. The station was abandoned, and one of the two 220/110kV, three-phase transformer banks was relocated to the Ertaishan substation and reconstructed as of May 2009. The Yinxing station includes 220kV and 110kV gas-insulated substation (GIS) equipment, dead-end and bus support structures composed of precast concrete tubes, and a two-story reinforced concrete moment frame with (presumably) URM infill control building. The principal high-voltage electrical equipment was furnished by Chinese manufacturers. We were told during our July 2009 visit that this substation had been constructed in 2006. All photos in this section show the substation as it was observed 14 months after the earthquake.

The 220/110kV transformer banks were supported on concrete grade beams and had welded anchorages to embedded steel. The high- and low-side bushings of the remaining bank appeared to be intact (Fig. 4.45), but surge arresters were damaged. Damaged 220kV bushings from the bank that had been removed were left in the yard, and cracked cast aluminum flanges were visible (Fig. 4.46).

The 220kV GIS equipment (Fig. 4.47) did not appear to be damaged, but it is unknown whether gas leaks or internal damage occurred. The 110kV GIS equipment (Fig. 4.48) was located at the back of the station against the slope and was destroyed in the landslide.

Wave traps were suspended from the dead-end structures with no lateral restraint; however, no damage conclusively attributed to this arrangement was observed. One of the wave traps was noted to be missing and may have fallen. All the surge arresters in Figure 4.49 were observed to have fractured, possibly caused by inertial overload on the surge arrestor directly or induced by conductor loading effects (nine of nine surge arrestors appear to have failed).

Fig. 4.44. Live tank breaker damage, Jiangyou Power Plant

Fig. 4.45. 220/110kV transformer

Fig. 4.46. 220kV bushing with cracked flange

Fig. 4.47. 220kV GIS equipment

Fig. 4.48. Tubular concrete dead end structure and 110kV GIS were damaged by the landslide

The landslide collapsed several A-frame type bus supports and dead-end structures at the rear of the station. The details of these structures are of some interest because of the failures observed in other stations, presumably from inertial or conductor pull forces. The precast concrete tubes have an outer diameter of about 40 cm (16 in.) with 6 cm (2.5 in.) wall thickness and are reinforced with 18 smooth reinforcing steel bars (about 12 mm [½ in.] in diameter) and a smooth wire spiral shear reinforcement (Fig. 4.50). The bars are anchored by welding to the interior wall of a steel sleeve at each end. To create a longer section of tube, two or more sections are joined together by welding one sleeve to another. Light-weight welded steel trusses, consisting of angle chords and bar web members, span the A-frames so that conductors or suspended equipment can be attached. The ends of the A-frames' tube columns appeared to be buried directly into soil with some concrete placed around the outer perimeter of the tubes.

The two-story control building shown in Figure 4.51 had significant damage to its stucco finish and portions of its infill. The infill consisted of concrete blocks and bricks in some places. The moment frame did not appear to be distressed. A single-story building at the site exhibited only minor damage to the stucco finishes. The perimeter brick with stucco finished substation walls had some cracking and spalling, but did not collapse.

Chuanxindian 110 kV Substation

Chuanxindian Substation is located near Muguaping village in a mountainous area about 50 km northwest of Deyang City and 8 km north of the town of Hongbai. The ASCE/TCLEE team visited the site 14 months after the earthquake. Along the road, substantial damage to one- and two-story URM buildings was observed. The team also observed structural damage and some collapses of buildings and silos at the cement and chemical manufacturing plant facilities along the river.

Fig. 4.49. Suspended wave traps, arrester damage (July 2009)

Fig. 4.50. Precast tubular concrete column reinforcement and end anchorage

Fig. 4.51. Control building damage

The station was de-energized and abandoned following the earthquake. It includes a 110/10kV, three-phase transformer (which has been relocated to another site), 110kV dead tank SF_6 circuit breakers, center-break disconnect switches on low-profile structures, PTs, arresters, unrestrained suspended wave traps seen at other stations, and dead end/bus support structures composed of precast concrete tubes. Except for the 110kV SF_6 circuit breakers, all of the high-voltage equipment was provided by Chinese manufacturers. Several small single-story concrete moment frame/masonry infill buildings are also on the site.

The substation perimeter walls (unreinforced brick) collapsed during the earthquake and had been rebuilt. The 110/10kV transformer was unanchored and had slid off of its foundation (Fig. 4.11). Additionally, some oil leaks were evident (Fig. 4.52). The substation buildings had only minor exterior damage; however, suspended lighting and ceiling tiles fell during the earthquake, as shown in Figure 4.53.

Utility representatives stated that the 110kV SF_6 dead tank circuit breakers were undamaged. This type of breaker is generally considered to be seismically rugged and is used in many high-seismic regions of the United States. No visible damage was observed in the disconnect switches or other structures, including the dead ends, which were constructed of the typical A-frame precast concrete tube columns. As shown in Figure 4.54, the A-frame structures used concrete beams instead of the light steel trusses. As seen in Figure 4.54, the potential for the hanging wave trap to swing apparently did not damage the bushing on the SF_6 circuit breaker.

Fig. 4.52. Unanchored transformer slid off this foundation, later relocated.

Fig. 4.53. Control building nonstructural damage

Fig. 4.54. Disconnect switch, bus supports, and SF_6 circuit breakers

4.3 Transmission Towers

Many high-voltage (110 kV and higher) transmission towers were partially or completely damaged in the earthquake. The predominant causes for tower failure were landslides that moved the tower foundations more than 3 m (in some cases more than 30 m) and localized damage due to rock falls. In one case, a tower collapsed apparently due to inertial loading, which is considered unusual (Tower #13, Guangyuan Electric Bureau). When towers collapse, there is the potential for adjacent towers to be pulled down. While we did not observe tower failures due to large debris flows that is not to say that they did not happen.

Landslides occurred frequently in the mountainous areas of Yingxiu, Anxian, Beichuan, and Jiangyou Counties. In these areas, PGA values commonly exceeded PGA = 0.3 g (sometimes as high as PGA = 0.9 g). We estimate that perhaps 1 in 100 towers along the approximately 200 km of roadway we observed were affected by landslides. With a common tower spacing of about 340 to 680 m (1,000 to 2,000 ft.) and distances between substations of about 16 km (10 miles), there are commonly 50 towers per circuit (110 kV) or 100 towers per 220 kV circuit. Thus, it is likely that one-third to three-quarters of all circuits had at least one tower failure.

Figures 4.55 to 4.58 show collapsed towers, apparently due to landslide. Figure 4.59 shows a transmission tower that remains standing despite debris flows on either side of its four legs, but the lattice steel arm on the ground use to be atop the tower. The damage to the arm might have resulted from inertial loads or pulldown loads.

Fig. 4.55. Collapsed tower

Fig. 4.56. Collapsed tower

Fig. 4.57. Collapsed tower

Fig. 4.58. Tilted transmission tower, possibly due to landslide (Yingxiu)

Fig. 4.59. Tower with damaged arm

Figure 4.60 shows a collapsed 220 kV suspension tower. PLA guards prevented us from climbing up to examining the foundation, but from viewing the overall transmission line, a landslide failure of an adjacent tower (not shown) might have resulted in pulling this tower down.

Figure 4.61 shows a reconstructed 220 kV tower, and Figure 4.62, a collapsed 110 kV tower. The two towers are located within 1 km of each other, adjacent to landslides on a mountain slope near Beichuan. We surmise that the 110 kV tower was pulled down by the landslide that affected the nearby tower. While the original tower locations remain unknown, most were placed on ridge crests instead of along slopes; undoubtedly, this reduced the total number of towers that would have been affected by landslides, as we observed many landslides to initiate just 1 m (about 3 ft.) beneath mountain ridges.

Figure 4.63 shows a collapsed low-voltage line that heads up the mountain. It is hard to see the tower poles in this photo; they were pulled down by a deep-seated landslide. The location is about 10 km west of Beichuan with estimated PGA = 0.3 g. The 220 kV towers at the ridge crest were unaffected by the landslide. Figures 4.64 through 4.70 show additional damaged transmission towers.

Fig. 4.60. Collapsed 220 kV transmission tower

Fig. 4.61. Reconstructed 220 kV tower

Fig. 4.62. Collapsed 110 kV tower

Fig. 4.63. Collapsed 10 kV line

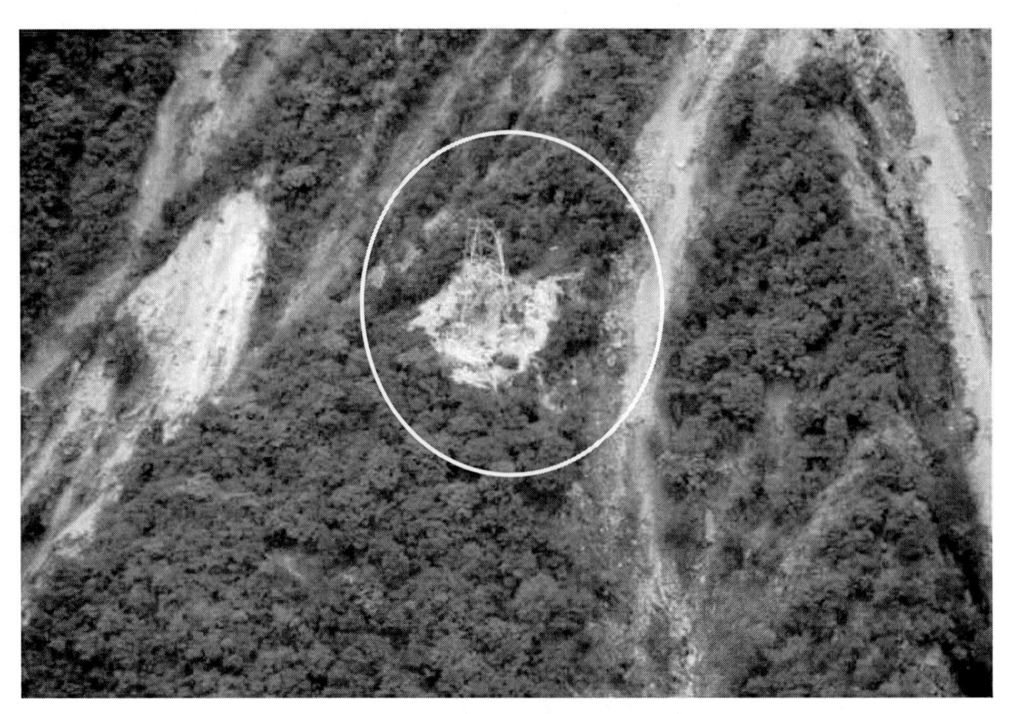

Fig. 4.64. Collapsed tower

Fig. 4.65. Collapsed tower, #123, 500 Maotan line

Fig. 4.66. Rack fall-impacted tower, #123, Hongxue line of Yingxiu Town

Fig. 4.67. Rock fall-impacted tower, #79, Zhouman line

Fig. 4.68. Collapsed Tower, #13, Jiangsang line

Fig. 4.69. Rockfall impacted tower, #14, Tianjin line, Jinlong Lake area, Wenchuan

Fig. 4.70. Collapsed tower, #11, Tianjin line, Jinlong Lake Area, Wenchuan

4.4 Distribution System

Overhead distribution (4 kV – 34.5 kV) is common in cities and towns in China. The most common pole style is a concrete tube. These concrete tubes appeared to have no shear steel (Fig. 4.71), and high bending/shear induced failures occurred where there were pulldown forces. Many dead-end frames at substations also use this construction, albeit oriented into an A-frame assembly. Several of these collapsed at a number of 110 kV and 220 kV substations with PGA > 0.4 g, (Fig. 4.30, 4.36, 4.48) due to a combination of inertial loading coupled with cable line dynamic loading (not landslide), leading to gross damage to yard equipment below.

Landslides and rock falls caused loss of poles in many areas (Fig. 4.72).

At Shenxigou village, we observed rockfalls that were likely triggered by the first rupture, followed by surface faulting of more than 3 m thrust upwards and 4 m lateral (some of this might have been induced landslide) that toppled power poles and caused power lines to come to rest atop the rock fall debris.

Platform-mounted transformers were observed. Figures 4.74 and 4.75 show two such examples. This type of damage was observed in a California earthquake in 1952, and since then, power distribution companies in California have all but eliminated the placement of unanchored transformers on elevated platforms for new installations. We recommend similar retrofits be adopted in China as a high priority in all high-seismic regions of the country.

Fig. 4.71. Collapsed 10 kV distribution pole (10-in. diameter, 2.5-in. wall thickness)

Fig. 4.72. Pull down of power and telecom distribution due to landside

Fig. 4.73. Surface faulting uplift of about 3 m and lateral shift of about 4 m with damaged road and tilted distribution poles

Fig. 4.74. Fallen distribution transformer

Fig. 4.75. Distribution system transformer (10 kV) being replaced (Hongkou), July 16 2008

Figure 4.76 shows the failure of a low-voltage (10 kV) power cable crossing a bridge at Dujiangyan. The movement of the bridge embankment by a few inches resulted in the non-ductile failure of the PVC pipe at two locations and one location at the other abutment. It is likely that the power remained in service because it appears that the cables did not break at the PVC failure points and there was no new power line observed.

Figure 4.77 shows a distribution transformer resting on a pile of recently assembled bricks. Its predecessor fell off this pole. PGA at this site was about 0.3 g.

The physical requirements for the generation, transmission, and distribution of electrical power in China result in systems that are similar to those throughout the world. The function of equipment and substation configurations are almost the same. The major differences we observed between China and recently built substations in California were usually associated with equipment installation details and the seismic design of building structures.

4.5 Lessons Learned and Recommendations

The high-voltage switchyard equipment damage to the China State Grid substations is the same type we have seen many times in earthquakes in the United States and around the world.

Fig. 4.76. Damage to low-voltage power conduit at bridge abutments

The damage at high-voltage substations was severe, resulting in significant power disruption. The damage was largely due to the collapse of live tank circuit breakers, failure of high-voltage transformer bushings, movement of unanchored high-voltage transformers, collapse of dead-end towers, collapse of substation control houses, swinging of suspended wave traps with limited to no cable slacks, and to a lesser extent, pulldown of disconnect switches due to failed live tank breakers. It is our understanding that essentially none of the equipment at the substations had been procured with any seismic requirements.

The large number of damaged control buildings was observed in the 2001 India Bhuj earthquake. The damage to building contents as a result of their failure contributed to system disruption and aggravated the overall power restoration effort. The reason these buildings did not conform to reasonably high seismic standard design codes for critical lifeline facilities is unclear, but should be considered a major flaw in seismic design philosophy in China.

It is vital that China institutes good construction and equipment installation practices for new facilities. Good installation practices involve adequate anchorage, which constitutes a very small percentage of total project cost for new construction. Adequate cable (or bus) slacks should be provided for connecting equipment. Improving inspection practices during the transition is also critical, as there is a tendency for simple tasks such as anchorage to be done without referring to construction plans. Construction plans must detail those elements that improve seismic performance. These include anchorage and providing adequate slack in conductors connecting equipment. New equipment (especially at 110 kV and higher voltages) should be procured with seismic requirements as specified in IEEE 693. Additionally, test requirements for pole mounted transformers and their installation designs should be developed to reduce distribution system failures or pole mounted transformers should be eliminated in the distribution system. The rebuilt Yingxiu, Ertaishan Substation showed a promising improvement in installation quality but lacked good seismic design.

Generally, retrofitting is not a cost-effective procedure in areas of moderate seismicity. However, three activities are generally recommended at all locations where the 475-year return period PGA exceeds 0.20 g.

- Station batteries should be restrained to their racks, and the racks should be of adequate design and be adequately anchored. Control cabinets should be anchored or otherwise adequately secured. Power transformers should be anchored, with anchorage forces computed using V = PGA * W (V is base shear, PGA is in g, W is total deadweight of the transformers including all oil and attached equipment) and the anchors designed for the corresponding overturning forces with a minimum factor of safety of 4 or more; the PGA value should be at least the 475-year return period value. When the PGA value is not known and the owner suspects the equipment is in a high seismicity region, use PGA = 0.5 g as a default.

- Live tank circuit breakers should be replaced for all critical circuits using either dead tank bulk oil (older style) or dead tank SF6 (newer style); all equipment must be adequately anchored. For areas of the country where the 475-year PGA exceeds 0.30 g, bushings for high-voltage transformers (200 kV and higher) should be seismically qualified (IEEE 396).

- Every dead-end tower in China within substations should be evaluated and, where necessary, rebuilt using materials (either steel or reinforced concrete) that can sustain the worst case loading of wind + ice + line loads or PGA=0.5 g without damage (keep stresses in steel below 0.96 F_y). All such structures that currently use under-reinforced concrete tubes should be replaced in a 5- to 10-year effort. Impact of the dead-end towers by adjacent under-designed items (like light poles) must be avoided; whenever possible relocate these items. The replacement of dead-end structures program should start with the critical nodes (substations) first.

Fig. 4.77. Unanchored distribution transformer

Fig. 4.78 Yingxiu, Ertaishan 220 kV substation rebuilt with one line working.

Fig. 4.79. The concrete platform with metal plates inserted for transformer anchoring shows one good seismic protection improvement

Fig. 4.80. The control room is built with a raised floor. It is not clear whether the equipment on the raised floor is anchored or not.

While costly, seismic improvement should be made for existing substation control buildings. Most if not all of the failures of the substation control houses could easily have been mitigated had China rigorously adopted and implemented a higher seismic standard for critical facilities. Upgrading substation control houses requires a long-term mitigation plan, but it is necessary because of the vulnerability of these buildings and the risks they present to the post earthquake performance mitigation program.

At many substations there are also residential buildings for substation workers. While these buildings do not house substation equipment, their collapse in the earthquake was prevalent, resulting in substantial life safety consequences. These buildings should be upgraded or replaced with suitably-designed buildings. Unreinforced masonry construction should not be allowed for human occupation where the PGA exceeds 0.15 g.

4.6 Acknowledgment

The information presented in this chapter was collected by several people over the course of several visits to the earthquake zone. Primary contributors are John Eidinger, Alex Tang, and Eric Fujisaki. Much information was provided by the Sichuan Electric Power Company, and their cooperation and courtesy is greatly appreciated. Some of the damage photos (courtesy of IEM) were taken the day or so after the earthquake in May 2008; others show the status of substations as of October 2008, May 2009, and July 2009, as discussed in the text.

We also appreciate Director Cui of Sichuan Association for Science & Technology, setting up a meeting and introducing the investigation team members to the engineering staffs of Sichuan Electric Power Company in Chengdu collecting valuable information to enrich the knowledge of lifeline earthquake engineering.

Unless otherwise specified in the figure caption, all photos are taken by the ASCE/TCLEE members.

4.7 Reference

Journal of Earthquake Engineering and Engineering Vibration, Vol 28 supplement, Oct 2008: General Introduction of Engineering Damage of Wenchuan Ms 8.0 Earthquake, Institute of Engineering Mechanics, China Earthquake Administration.

5 POTABLE WATER SYSTEMS

OVERVIEW OF THE EXISTING WATER SYSTEMS

The earthquake affected a large area of Sichuan Province that includes several major cities and a great number of towns and villages. There are mainly two types of water supply systems. In the major cities, water is supplied by water works companies that are typically owned by the government but independently operated. In the countryside and remote areas, water sources are mostly spring water and wells. The wells and spring water collection systems are either operated by townships, village governments, or by a group of homeowners.

Water used for drinking is always boiled (before or after the earthquake)—even in the cities—due to the uncertain sanitary condition of the distribution systems. We observed open sanitary sewers leading directly into creeks. We did not observe any water treatment (chemical disinfection or flocculation/sedimentation) at water intakes in mountain villages.

In the Chengdu Metropolitan area, water facilities were damaged in many locations, but the water service in small townships and rural villages were more severely affected by the earthquake. A water shortage after the earthquake affecting about 690,000 people was reported in the Chengdu metropolitan area, which has a population of more than 10 million.

In a Dujiangyan suburb, the peak ground acceleration (PGA) likely ranged from 0.20 g to 0.35 g, water outages were 5 days, power outages were 10 days, and natural gas outages exceeded one month according to a restaurant owner.

The China Urban Water Association reported that 7,800 km of water pipes were out of service, and 839 tanks were damaged. During the ASCE field investigations, we were unable to confirm these numbers, but it was readily apparent that the kilometers of pipes out of service were likely due to the assumption that all pipes in various mountainous towns and villages were out of service for one reason or another. We did observe a few damaged water tanks, but nothing like the 839 tanks reported with damage.

5.1 Water Pipes

Based on available information as of October 2008, we estimate that 30,000 km of buried water pipelines were subjected to reasonably strong ground motions (about PGA 0.10 g or higher). Service was restored to about 21,880 km of pipeline 20 days after the earthquake. Most of the out-of-service pipes appeared to be in areas that experienced PGA > 0.25 g. The causes for the service disruption include:

- damage to buried pipes due to landslide (prevalent), strong ground shaking (prevalent), differential movement at bridge abutments (occasional), fault offset (limited), housing collapse affecting service laterals (very common), and liquefaction (infrequent in the mountainous areas, possibly a major factor in the cities in the Sichuan plain);

- lack of manpower and parts to make repairs; and
- relocation of people from structures meaning that restoring buried pipe service to collapsed structures was not a priority.

This suggests that through the entire region perhaps as many as 2.9 million people were without piped water 20 days after the earthquake.

When we visited in October 2008 (160 days after the earthquake), new water pipes and supplies had been installed in towns and villages at all locations that had not been abandoned entirely. The entire town of Beichuan (population exceeding 50,000) was abandoned, so there was no effort to restore water supply. In many other towns and villages where there were remaining populations, about 70 percent or more were living in post-earthquake constructed housing; this housing was served by brand new water pipes laid on the surface. The larger new pipes were fusion-welded, high-density polyethylene (HDPE) (mostly 6–in. diameter mains), with branch pipes (mostly 1-in. polyvinyl chloride [PVC]) leading to the front doors of the temporary housing. We observed just one fire hydrant, located next to a new low-voltage power substation. In mountain villages, new water pipes consisted of 1-in. diameter PVC or rubber hoses leading form local streams downhill to single family dwellings.

Based on available models, the damage to buried water pipe (primarily asbestos cement, cast iron and PVC, all with push on joints and assuming 9,000 km of pipe exposed to PGV = 30 inches per second, on average) would be about 1,325[1] pipe repairs due to shaking.

In large landslide areas, existing buried pipe were destroyed over the width of the landslide zone (commonly 50 to 200 ft. wide), and as a rough estimate, there might have been 300 of these instances throughout the earthquake zone. The team observed several of them. Local surface faulting broke some pipes (about 50 to 100 repairs), and liquefaction damaged some pipes[2] (about 200 to 300 repairs), therefore the first-order-magnitude number of water pipe repairs would be about 1,925 to 2,225. The damage to service laterals of collapsed buildings in Mianzhu alone were reported to be 700; thus, the region-wide damage to service laterals would be several thousand. In towns such as Beichuan that were subjected to massive landslides, the number of broken water pipes remains unknown.

As the vast majority of residential buildings in the mountainous areas were either partially or completely damaged, there was little evidence of any attempt to repair the original water pipes. Instead, the strategy was to build new temporary shelter buildings, and then build a completely new water pipe distribution system to these shelters.

---

[1] The equation of repairs is 0.8 * 0.00187 * 30 * (9,000 * 3.28).

[2] This hazard is relatively uncommon in the mountains, but it is important in Mianzhu and other cities in the Sichuan plain.

In larger cities, such as Dujiangyan, where damage to buildings was sporadic, repairs were made to individual buried pipes, with new pipes limited to evacuation zones with newly-installed above ground water pipes.

In Chengdu, where few buildings were damaged, pipeline damage was also rare.

As of this writing, we have not been able to obtain accurate damage reports from local public works agencies, so the estimates provided are more of a forecast tempered by limited field observations covering perhaps 35 percent of the populated damaged area. However, from the data we have been able to obtain from the China Earthquake Administration (CEA) and from local interviews, it is clear that pipeline repairs were not uncommon.

Fig. 5.1. PVC water pipe with split

Fig. 5.2. Water leak under road

These pipeline damage estimates exclude damage to service laterals from the collapse of local buildings. This type of damage, which often has to be repaired by the building owners, has been extensive when there are many collapsed structures, as it was in the 1906 earthquake in San Francisco.

Fig. 5.3. Repair to water and sewer pipes in Dujiangyan

Fig. 5.4. Broken water pipe in Chengdu City

Fig. 5.5. Broken water pipes at either side of a landslide, Shenxigou Village (near Hongkou); landslides cause more damage than earthquakes.

Fig. 5.6. Damage mechanisms to water pipes at one side of a landslide, Shenxigou Village

Figure 5.7 shows a 6-in. diameter, fusion-welded HDPE pipe in Yingxiu 160 days after the earthquake. As essentially all of the residential buildings were destroyed or rendered uninhabitable in this town, the remaining population was relocated into temporary shelter buildings, as seen in the background in Figure 5.8. All of these buildings had branch pipes leading to them, allowing residents to obtain water via outdoor hose bibs; it is unclear if cold weather in winter will lead to pipe freeze.

Fig. 5.7. Temporary above ground water pipe in Yingxiu for temporary shelters

Fig. 5.8. Temporary above ground water pipe with hose bib to temporary housing in Yingxiu

In Pengzhou City, about 15 km north of Chengdu, there were 59 water pipe repairs (perhaps PGA = 0.10 g to 0.15 g ground motions).

In Anxian County, there were 100 pipe repairs for 39.6 km of pipe mains (70 percent cast iron, 30 percent steel), coupled with 700 repairs to service laterals (polyethylene), mostly to buildings that had collapsed.

Fig. 5.9. Damaged cast iron pipe, Mianzhu

Fig. 5.10. Damaged cast iron pipe, Guangyuan City

With a dedicated pipe repair crew of 100 people, we would expect that about 40 pipes could be repaired each day. This would suggest a total pipeline repair time of perhaps 50 days or so. However, as of October 20, 2008 (about 160 days post-earthquake), it was evident that underground piped water had been restored only to the cities in Sichuan plain (PGA under 0.25 g), with above ground temporary piped water being available to almost all emergency shelter areas in the mountain villages and towns (PGA over 0.30 g).

It was reported that by June 14, 2008, Chengdu Waterworks had installed 6,500 m of temporary piping for distributing water to center locations of the temporary housing, or tent encampments.

5.2 Water Facilities

One report described damage to water treatment plants with structural cracks in the building walls; however, the plants apparently remained in operation. Pumping plant building structures, presumed to be an unreinforced masonry (URM) structure, appeared to be heavily damaged; however, the pumps were typically still functional. Crews installed steel pipe bollards to protect the pumps from the falling debris of the surrounding structures.

The headquarters of the Chengdu Waterworks sustained some damage, and the control room was temporarily relocated from the fifth floor to the ground floor on May 20, 2008 in response to the cracked walls and aftershocks. The control room was returned to the original location on June 20, 2008.

By May 31, 2008, 4,080 wells and water supply points were reported to be back in service (about two-thirds of the total). In towns and villages in the mountainous region, we observed water supply points as small diversion works from mountain streams, commonly built using rocks and mortar, leading into gravity-fed distribution pipes.

From our observation, more than than 95 percent of the elevated cast-in-place concrete water tanks performed well and remained functional (leak tight). These were located in areas with estimated PGA < 0.30 g. However, some of the water towers (assumed to be URM) sustained heavy damage, ranging from outright collapse (Fig. 5.11) to severe damage (Fig. 5.12, large x-cracks in the supporting structures) to no apparent damage (Fig. 5.13). We did not observe this style of tank construction in the mountain towns and villages, so it is uncertain how the tanks would have fared when subjected to PGA > 0.4 g. In the mountain town of Yingxiu (PGA between 0.4 g to 0.7 g), a rectangular concrete tank failed (Fig. 5.15).

The local water supply distribution system in some areas is unable to keep up with peak water demands, which is not uncommon in developing nations. In some smaller towns, some individual buildings have their own wells and small roof-top storage tanks. These owner-installed water tanks are filled with local well water (or town-supplied water when system-demand is low), and then used for building-specific purposes. One such tank (Fig. 5.14) survived, although supported atop a collapsed brick building.

Fig. 5.11. Collapsed elevated concrete water tank

Fig. 5.12. Elevated concrete water tank with heavy damage to tower

Fig. 5.13. Water tower in Dujiangyan, no apparent damage from exterior view

Fig. 5.14. Local storage tank atop damaged URM structure (Shifang)

Figure 5.15 shows a damaged reinforced concrete rectangular water tank. The roof of the tank was unanchored and composed of precast concrete slab elements; several of these slab elements slid sideways (at Yingxiu, ground motions likely exceeded 0.4 g) with some dropping into the tank. There had been no attempt to repair the tank 160 days after the earthquake.

Fig. 5.15. Local storage tank (Yingxiu)

Fig. 5.16. Storage building (Yingxiu)

Figure 5.16 shows the ruins of the wood post-and-beam maintenance facility with infill brick walls of a water and local power distribution company, which was located adjacent to the water tank (Fig. 5.15).

Figure 5.17 shows a rock-and-cement lined storage basin that was used to store fish in Shenxigou Village. It experienced about a 1-ft. vertical and a 1.5-ft. horizontal surface fault offset through the far wall. The basin remains unrepaired and empty more than five months after the earthquake, reflecting the impact of the earthquake on the local village economy, which is primarily tourism.

We noted that the style of wall, with the mortar-cemented stonework, remained mostly intact so that the release rate would have been modest. We did not observe gross downstream erosion; the bulk of the water would have been released into a small creek. Also notable is that 160 days after the earthquake neither reservoir was restored to service. This reservoir was used as fish-holding ponds, demonstrating that the economic activity in the area was still affected by the quake.

Fig. 5.17. URM water storage basin, Shenxigou Village

5.3 Water Emergency Response and Recovery

In a major disaster with extensive destruction of homes, water system recovery includes two progressive steps:

1. To provide water to central locations and temporary housing locations, such as disaster control centers and tent encampments; and
2. To repair and recover the existing water systems.

The water works companies and the military jointly handled the first step and appeared to have performed well as evidenced by the relative calm of the many displaced residents. Temporary water treatment facilities were installed near the temporary housing to supply water to the distribution centers. In most small villages and towns in the mountainous areas, we also observed above ground pipe (fusion-bonded HDPE, rubber hose, or cement-jointed PVC) laid to original houses (estimated under 20 percent) that were not damaged to the extent that they had to be evacuated or demolished. Waterworks companies also distributed water disinfection pills throughout the towns and villages in the remote areas. The average post-earthquake usage for per person is about 30 liters per day.

As of October 21, 2008, 160 days after the earthquake, we observed essentially no attempt to restore water via the original buried pipe system in the towns and villages in the mountains (PGA>0.4 g). Largely, we attribute this to the near total destruction of original residences and many industrial buildings; thus, there was no need for the immediate reconstruction effort.

Additionally two days after the earthquake, a rumor spread rapidly throughout Chengdu City that the water has been contaminated by an explosion in a chemical factory at the water intake. The company received more than 1,900 phone calls that day compared to average 200 calls on normal days. Despite the company's denial, the rumor created a panic, and the city's residents started storing water at home with pots and pans and rushed to buy bottled water. The pressure of the water system suddenly dropped to a dangerously low level. To avoid the damage from the negative pressure from the heavy water usage, the water system was shut down for a short period. The water agencies and the city government had to start a public campaign with all available means, including cell phone text messaging, to refute the rumor. The water pressure was returned to normal that afternoon.

Figure 5.18 shows newly installed above ground water pipes in Dujiangyan. These pipes were installed to temporarily deliver water to nearby temporary housing.

The military also set up many sites with a reverse osmosis system and bladders to provide emergency water supply to local communities (Fig. 5.19).

Fig. 5.18. New above ground water pipes (Dujiangyan)

Fig. 5.19. Bladders set up by the People Liberation Army to provide water to local communities

5.4 Dujiangyan Water Diversion System

In about 256 BC, Li Bing, the governor of what was then the Kingdom of Qin, and his son created a flood regulation system in Dujiangyan along the Minjiang (Min River). The area of the original works currently includes a series of check dams, flood diversion works, and a public park land including many wooden temples dedicated to Li Bing and son (Fig. 5.20).

This area was exposed to ground shaking of perhaps PGA = 0.25 g. One of the wooden temples collapsed; others were lightly to heavily damaged.

Figure 5.21 shows repairs to the embankments (background), denoted by the light-colored material. The cracks at this location were commonly 2 to 4 in. wide, running over a few thousand feet of river side. Higher up the embankment, there were clearly observed settlements of 2.5 to 5.0 cm (1 to 2 in.) and lateral spreads of 2.5 to 5.0 cm (1 to 2 in.).

Fig. 5.20. Monument of Li Bing and son in Dujiangyan

Figure 5.22 shows the remnants of broken masonry railings at this site. Based on back calculation of the ground acceleration needed to cause bending moment-induced failures, we estimate that the site experienced between PGA = 0.25 to 0.50 g. Replacement masonry decorative railing walls were being installed on October 17, 2008, with identical design (they will fail again in a future similar-magnitude earthquake).

Figure 5.23 shows the embankment protection downstream of the check dam with repairs.

Tall under-reinforced (or unreinforced) monuments toppled at this site (Fig. 5.24). Note the steel pull out and concrete failure for the two rebar-held connections for the bottom portion and the non-reinforced connection for the top portion. Upon toppling, the top of the monument snapped.

Figure 5.25 shows two wooden temples built to honor Li Bing. On the left side, the top two floors collapsed. These temples were closed to the public due to the damage. Because this site is a major tourist attraction, this hurts the local economy due to a reduction in tourism and a loss of a cultural artifact representing one of the most famous engineering flood control achievements in Chinese history.

Fig. 5.21. Repair of embankment (background) and base of collapsed masonry railing wall (foreground)

Fig. 5.22. Broken railing wall, note the posts snapped at the base

5.5 Lessons Learned and Recommendations

The lessons learned from this earthquake on the performance of the water system include the following:

- Elevated reinforced concrete tanks, when designed to reasonable seismic standards, performed reasonably well.
- Non-engineered unreinforced masonry structures performed poorly, and many collapsed.
- Damage to pipelines where they cross zones subject to landslide performed extremely poorly. Repairs to pipelines at landslide zones generally require installing entire lengths of new pipe.
- Dislocations of a large number of people from houses with structural damage led to an initial large effort to provide water via tanker truck and, subsequently, via above-ground pipes to temporary housing camps. Delivery of the quantity of water needed and monitoring the water quality is resources intensive.
- Restoration of the water supply via pipes was largely complete as of five months after the earthquake (99.5 percent+).

Fig. 5.23. Repairs made to downstream slope protection

Based on the team's findings, the following are suitable recommendations for water utilities located in high-seismic zones:

- All structures (water tanks, pump station buildings) should have suitable seismic design. In areas subject to ground shaking much over PGA = 0.3 g, a minimum seismic design should be for base shear V = 0.14W, with ductile design details. Higher seismic design standards will provide for better performance of the structures. Unreinforced masonry infill walls are not acceptable.

Fig. 5.24. Toppled monument

Fig. 5.25. Collapsed and heavily damaged temple of Li Bing and his son.

- A high percentage (more than 90 percent) of engineered reinforced concrete tanks performed satisfactorily in this earthquake, defined as tanks that remained leak tight.
- Water utilities should prepare suitable surface or near surface geology maps that identify pipelines that traverse zones subject to liquefaction, landslide, or fault rupture. Pipelines that traverse these zones should use materials that have at least some seismic capacity to sustain ground deformations.
- Transmission pipelines should avoid traversing steep landslide zones if possible; it is not feasible to design buried water pipelines to remain in service given landslide movements on the order of 10 to 100 ft.

5.6 Acknowledgment

The authors appreciate the information provided by the China Earthquake Administration (CEA) of Beijing and the data from the Institute of Engineering Mechanics (IEM) of Harbin, China.

All photos used in this chapter were taken by the ASCE/TCLEE earthquake investigation teams, unless otherwise specified in the caption of each figure.

6 TELECOMMUNICATIONS

EXECUTIVE SUMMARY

Three factors affected the performance of the telecommunication systems in the earthquake-devastated areas. The first and most common was the increased call volume, which impaired the network. In a post disaster environment, call volume usually doubles or triples, but systems are typically designed for their daily highest traffic rate. The second factor was the long duration of electric power outage in most remote areas. Some areas were out of power for as long as 20 days. Lastly, there was damage to telecommunication equipment buildings, switching offices, cell sites, equipment, and cellular towers. One of the major service providers in Sichuan suffered damage to more than 600 equipment buildings and 6,000 km of optical fiber cables.

In the remote area of Hongkou, one cellular service provider was not available during our July 27 field investigation. Cellular service was the most affected telecommunication service in the affected areas. The general outage was between three and 10 days, with some exceptions. As the ASCE team did not have an opportunity to meet with any of the service providers in Sichuan, more accurate data will be required to provide a concise analysis of the telecommunication service performance. Any additional information that becomes available at a later date will be posted on the TCLEE/ASCE Web page.

The acute demand of service and the high-velocity deployment process were the main reasons that the system was not constructed with the precautions needed for a more robust system. However, this event provides Sichuan Province an opportunity to rebuild its system incorporating resiliency in the process in addition to upgrades.

6.1 Description of System

China currently has four major players in the telecommunication service industry. They are China Telecom, China Netcom, China Mobile, and China Unicom. These companies are all publicly traded, with China Telecom listed on the New York Stock Exchange. Prior to 2002, China Telecom was a state-owned enterprise. China Netcom is a spin off from China Telecom, which owns almost all assets in southern China. Both China Mobile and China Unicom only provide cellular service. China Mobile uses Global System for Mobile Communications (GSM) technology, originally Groupe Spécial Mobile, while China Unicom uses Code Division Multiple Access (CDMA) technology. Both companies are deploying third generation (3G) technology to provide customers with additional function including video streaming. 3G technology converges GSM and CDMA into a common platform; therefore, handsets will be simplified.

The network configurations for both landlines and cellular lines are similar to that of the North America network. Due to timing, most of the networks in China were deployed with fiber-optic cables.

In the earthquake affected areas, landlines were not that common, except in the bigger cities. Most of the rural towns depend on cellular services as it is faster and more effective to deploy cellular service in a mountainous terrain.

6.2 Performance of the Cellular (Wireless) Network

The cellular network configuration in China is identical to that found in North America and Europe. The components—the technology used in network interconnection, and the application specific firmware and software—are also similar. The network consists of two major facilities, cell sites (also known as base stations), and mobile switching offices (also known as mobile telephone switching office). The cell sites are interconnected as well as connected to the mobile switching offices. The equipment used in these facilities varies depending on the size, location, and function of a particular site or office. Because cellular customers are mobile whether local or roaming, one or more facilities are required to store the customers' data to allow billing as well as connecting the handset (cellular phone) to the network.

Cell sites usually consist of a tower for mounting the antenna, a small building to house radio equipment, power equipment (batteries and rectifiers), a base station controller, transmission equipment, and environmental control equipment. Mobile switching offices consist of switching equipment, power equipment (batteries and rectifiers), transmission equipment, data storage and processing equipment, and environmental control equipment. The mobile switching offices usually have a power generator installed as the emergency power source. The interconnection between cell sites and mobile switching offices uses landline (copper or optical fiber cables) and/or microwave.

The cellular network in the earthquake-affected areas sustained damage to the following components:

- tower;
- building;
- equipment;
- cables, which are often severed between cell sites and mobile switching offices;
- utility poles; and
- power source.

The service interruption ranged from three days to five weeks, depending on the areas within the earthquake-impacted area. The interruption in cities seemed to be much shorter than that in rural areas. The main reason was access to damaged sites due to the high quantity and size of landslides. The People Liberation Army (PLA) of China played a key role in clearing debris and rocks to provide access for various emergency service groups.

Hongkou Cell Site

Two cellular service providers, China Telecom and China Mobile, were colocated about 40 to 50 m from the west side of the road to Hongkou. The equipment buildings were adjacent to each other (Fig. 6.1). There were no microwave dishes on the towers to interconnect with another cell site or mobile switching office; connection was by means of cable. Most likely optical fiber cable was used as damaged optical fiber cables were observed within 20 km in this site.

Both buildings and the equipment inside them sustained extensive damage. It was noted that the equipment was not anchored. In addition to the telecommunication equipment and batteries, the air conditioning units were also damaged (Fig. 6.2 to 6.5). A portable power generator indicated that power was out at this site (Fig. 6.6). A 10-kV pole mounted transformer not far from this site was damaged and was being replaced during the team's July 17 visit (Fig. 6.7). China Telecom built a new building to house the equipment (Fig. 6.8).

Fig. 6.1. Hongkou cell sites: the taller tower belongs to China Mobile, the other to China Telecom.

Fig. 6.2. The unreinforced masonry (URM) building was damaged and almost collapsed

Fig. 6.3. New equipment was temporarily installed to restore service

Fig. 6.4. Damaged batteries due to building damage

Fig. 6.5. Air-conditioning units were damaged; note the building debris in the foreground

Fig. 6.6. A portable power generator was used to power the cell sites after emergency recovery at Hongkou. This photo was taken in front of the damaged cell site building on the pathway leading to the main road.

Fig. 6.7. The damaged 10-kV pole mounted transformer was being replaced during our site visit on July 17, 2008. Note the location of the cell site with towers showing.

Fig. 6.8. The new cell site building was built on a new concrete pad.

Yingxiu Cell Site

Yingxiu was one of the hardest hit communities in this earthquake. The town was almost flattened. During our site visit in October, most of the collapsed buildings remained untouched. The essential lifelines were back in service. A temporary power substation was built to provide power to the survivors and the emergency shelters. The China Mobil cell site is about 120 m from the temporary power substation. Figure 6.9 is the cell site in Yingxiu that the ASCE/TCLEE team visited. Note that the door was not installed in this new wall. Figure 6.10 shows some of the bricks from the old damaged wall laying around one of the foundations of the tower inside the wall.

Fig. 6.9. A China Mobile cell site in Yingxiu. Insert shows a missing door in the new wall.

Fig. 6.10. Bricks from the damaged wall all over the yard under the tower

The local people told us that some of the antennae had dropped and the top part of the tower was damaged. The brick wall around the cell site collapsed and was rebuilt during our visit in October. The equipment building inside the brick wall seemed to be quite new with no obvious signs of damage. We believe that repairs were made prior to installing new equipment to restore this cell site.

Inside the building, the equipment and the cable routing hardware also seemed new (most likely replacing the damaged equipment) but was unanchored. A new power generator was also installed in this building. It was not operating during our visit as the power was back on in the area. Figures 6.11 to 6.16 show the new equipment and hardware in the building.

In addition to fiber cable connecting the cell site with switching offices and other cell sites, a temporary satellite dish was installed to provide connections with other remote locations (Fig. 6.17).

Another cell site that is located behind the collapsed high school in Yingxiu was leaning about 10° from vertical (Fig. 6.18). We were not able to get close to this cell site due to the collapsed buildings around it.

Fig. 6.11. Power equipment in the cell site building

Fig. 6.12. Radio, controller, and transport equipment look quite new

Fig. 6.13. New overhead cable racks

Fig. 6.14 New cable rack above the base station controllers

Shifang Cell Site

Most of the chemical factories are located in this area. Extensive damage to factories and buildings was observed. The degree of damage in this area was equal or even greater than in Yingxiu and covered a much wider area.

The PLA performing the emergency services in Shifang set up a number of small cell sites to relieve the telecommunication failures. The service interruption in Shifang was up to two weeks.

Fig. 6.15. This new power generator for the cell site was not anchored. The exhaust pipe was not connected to the outside.

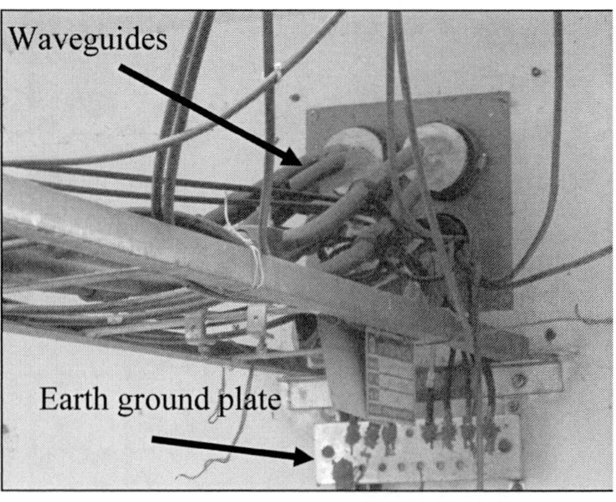

Fig. 6.16. The earth ground plate at the wave-guide entry to the building looked new, and the cable rack wall bracket was new.

Fig. 6.17. A temporary satellite dish was used to communicate with other nodes in the network. Note that there were two portable power generators and a new air conditioning unit on the outside of the equipment building. This equipment was brought in after the building was repaired.

Fig. 6.18. Cell site antennae tower (circled) slanting about 10° from its vertical position. It might be leaning against one of the collapsed buildings.

Figures 6.19 and 6.20 show temporary cell sites set up to bring telecommunication service back to the area.

The China Earthquake Administration (CEA) team reported that the cell sites at Jianshi, Wudu, and Mianchi sustained service interruption damages, and the cell tower at Mianchi collapsed (Journal of Earthquake Engineering and Engineering Vibration [JEEEV] 2008).

Fig. 6.19. One of the temporary cell sites set up to provide service in Shifang

Fig. 6.20. A temporary cell site set up on the military campground

Fig. 6.21. A temporary cell site set up on the side of a main road in Dujiangyan

Mianzhu Cell Site

A roof-top cell site on an apartment building in Mianzhu was also damaged (JEEEV 2008).

Dujiangyan Cell Site

Dujiangyan, one of the big cities within the earthquake impacted area, is within 60 km northeast of the epicenter. Within the city core, many buildings were damaged, partially collapsed, or collapsed. A few cell sites were damaged. The CEA team reported that a switching office building and equipment were also damaged.

In passing by this city, we noted a number of temporary cell sites set up on the sidewalk to provide service (Fig. 6.21). These will be removed once the sites damaged in the earthquake are repaired and restored to service.

6.3 Performance of Cell Sites and Switching Offices Reported by IEM (Institute of Engineering Mechanics), Harbin

Zifa Wang of IEM and his colleagues provided the following information and permission to use the photos in this section.

A team of earthquake research engineers from IEM and many other universities within China performed a field investigation within two weeks of the earthquake. The team reported on all earthquake damage including lifelines. Their goal was to capture the perishable information for indepth study to understand the modes of failure. Due to extensive damage, it was difficult and time consuming to catalogue each item. The date collected will help earthquake engineers mitigate future damage and save lives.

Cell Sites

Figures 6.22 to 6.31 show various modes of cell site failure in the earthquake-impacted area.

Switching Offices

Figures 6.32 to 6.34 show damage observed by the IEM team.

Fig. 6.22. The walls and part of the roof of this cell site in Mianzhu on the top floor of this commercial building collapsed and damaged the equipment inside. (Courtesy of IEM)

Fig. 6.23. A damaged cell site in Dujiangyan (Courtesy of IEM)

Fig. 6.24. This cell site building in Jianshi, Shifang, collapsed and damaged all the equipment inside. (Courtesy of IEM)

Fig. 6.25. The roof of this cell site building in Wudu, Shifang, collapsed on the equipment inside. (Courtesy of IEM)

Fig. 6.26. The concrete floor of this cell site in a metal panel building in Dujiangyan cracked causing equipment failures. (Courtesy of IEM)

Fig. 6.27. The battery bank in the same building escaped damage, but the ground plate was pulled from its mounting location. (Courtesy of IEM)

Fig. 6.28. The walls of this cell site building in Dujiangyan collapsed and damaged the equipment. (Courtesy of IEM)

Fig. 6.29. The cables in the other room of the building came off the cable rack above. The collapsed wall is on the left in this photo. (Courtesy of IEM)

Fig. 6.30. A collapsed cell tower in Mianchi, Shifang. This tower was damaged by the rock fall. (Courtesy of IEM)

Fig. 6.31. The walls of this cell site building collapsed damaging the equipment inside. (Courtesy of IEM)

Fig. 6.32. Inside a cell site administrative office, book cases and file cabinets and documents were scattered on the floor. (Courtesy of IEM)

Fig. 6.33. Battery damage caused by fallen cabinets in a switching office in Dujiangyan (Courtesy of IEM)

Fig. 6.34. Damaged air conditioning units. The small equipment fell from the top of the units onto the floor. (Courtesy of IEM)

6.4 Performance of the Landline Network

This section includes observed damage to switching offices. Most likely the landline network sustained the worst damage of the telecommunication system in the earthquake areas. Massive landslides and rock falls along with the number of damaged bridges and buildings contributed to the landline failures.

The failure of cell sites, switching offices, and landlines connecting these nodes as well as network traffic congestion were the main reasons for poor telecommunication performance during the first few days after the earthquake. The following presents our observation of landline damages during the investigation trips to the earthquake-impacted areas.

Road to Yingxiu

Figures 6.35 to 6.36 are a few samples of the observed damage to aerial cables on utility poles. With the recent deployment of optical fiber cables to scale up the capacity of the network to deliver both voice and data, all the damage observed were optical fiber cables.

A telephone booth at Yingxiu was totally damaged, leaving a sign on the side of the road (Fig. 6.38).

Fig. 6.35. Damaged utility pole on the road to Yingxiu and a close look at the cables, which are all optic fiber cables.

Fig. 6.36. Damaged pole due to rock falls

An optical fiber cable was installed along the road leading to Yingxiu for emergency response purposes (Fig. 6.37).

Fig. 6.37. Sign showing a China Mobile emergency response optical fiber cable along the side of the road leading to Yingxiu.

Fig. 6.38. A public telephone booth sign left on the side of the road near the damaged phone booth

Fig. 6.39. A cross connect box mounted between the utility poles was opened to allow the service crew to remove connections and connect telecommunication service to temporary shelter areas.

Due to the large number of building collapses, the cross connect box for the neighborhood was opened to allow the service crew to remove unwanted connections (Fig. 6.39). The strong shaking in this area did not damage poles. Most likely the platform joining the two poles together provided added strength and also it is not in the landslide area.

Road to Hongkou

The failure mode of the poles in this area is quite different from that observed in Yingxiu. A number of poles here broke at about one-third of their height from the ground. Although some of the broken poles resulted from fallen rocks or trees, a few were broken by very strong shaking. The surface faulting in this area provided us with a sense of the ground shaking intensity during the earthquake. This popular vacation area is about 15 to 20 km from the epicenter and is in a valley between two mountains.

Figures 6.40 to 6.43 provide a collection of failure modes observed.

Fig. 6.40. A pole damaged by the rock falling from the slope across the road

Fig. 6.41. A pole damaged by the rock fall, note that pole on the left was not damaged

Fig. 6.42. Both concrete and wood poles were damaged at this section of the road leading to Hongkou. Fallen rocks were pushed to both sides of the road to clear the road for cars.

Colocating lifelines creates collateral damage. When this bridge span collapsed (Fig. 6.44), the telecommunication cable alongside was also damaged.

Road to Nanba and Hejiaba

While Nanba (Mark #4 in Figure 6.1) is located about 120 km from the epicenter, it is another hard hit area where a large number of buildings collapsed. The two bridges in town were damaged. The observed pole damage along the road leading to Nanba was quite similar to those shown in the section above. The intent is to show the extent of damage miles from the epicenter. This also underscores the high demand of resources to handle emergency response with a system that sustained so much damage.

Figures 6.47 to 6.51 show the damage to the landline network and the temporary cable connections to provide service to these two towns.

Poles were shipped to Hongkou to repair the damage (Fig. 6.45). Figure 6.46 shows the repair crew was installing new poles for running the cable.

Fig. 6.43. There was no rock fall and no landslide in this neighborhood. Most likely this pole was damaged by the strong shaking as surface fault was observed in this area.

Fig. 6.44. The telecommunication cable on the right side of the collapsed span was damaged.

Fig. 6.45. New poles shipped to Hongkou

Fig. 6.46. The repair crew installing poles to replace temporary cable

Fig. 6.47. Extensive landslides along the steep slopes of riverbank toppled many utility poles

Fig. 6.48. Utility pole damaged by collapsing building

Fig. 6.49. Temporary repair of the landline with splice case tied to the pole by a piece of cable. Note the temporary power cables also shared the same pole here in Hejiaba.

Fig. 6.50. The power lines on the top portion of this pole were not energized. Note the temporary telecommunication cables and the splice case mounted at about mid-height of this pole.

Fig. 6.51. This shows a temporary optical fiber cable bringing the connection from Nanba (right side) to Hejiaba.

Fig. 6.52. This utility pole was damaged, and the cables were removed. On the left of the pole in the background is a chemical factor that sustained extensive damage.

Road to Shifang

Shifang (Mark #3 in Figure 6.1) is an area with many chemical factories. Damage to structures and lifelines in this area were quite severe. Within the area's collapsed residential buildings, many cell sites and switching offices sustained various degrees of damage. Many temporary mobile roadside cell units were set up by the PLA.

Figure 6.52 shows a utility pole slanting about 35° from its vertical position due to ground movement and potentially shallow embedment. In the background is a chemical factor producing fertilizer. This factory sustained extensive damaged to its old structures.

A pair gain unit mounted between two utility poles was damaged by a piece of fallen concrete from the nearby building (Fig. 6.53). Fortunately, the unit did not fall off its mount.

Beichuan Area

This town (Mark #6 in Figure 6.1) was totally demolished by the earthquake, landslides, rock falls, and debris flow after the earthquake due to heavy rain. This town suffered the most casualties. Although there are no longer residents in the town, telecommunication is still a problem. The town has a number of switching offices and cell sites that link to the overall network of China. Bypasses must be carefully planned and executed to retain the efficiency and capability of the overall network in the province. Then decision whether to rebuild the town or to move it nearby will affect the plans to rebuild the network.

Fig. 6.53. Note the concrete slab on top of the pair gain unit on the platform between the poles.

Fig. 6.54. Due to landslide a pole was hanging by the cables along the edge of the slope failure.

Fig. 6.55 A pole collapsed along the slope and broke into four pieces.

Fig. 6.56. A splice case for a temporary optical fiber cables was placed on the side of the road.

Fig. 6.57. Severed optical fiber cables were left on the side of a road. Note the size of the rock that rolled down the slope above, some cable was under it.

Fig. 6.58. The temporary emergency cable was wrapped around the pole with a very tight bend radius.

Fig. 6.59. The temporary emergency cable was hanging quite low, with the splice case supported by the cable instead of a guide wire between the poles.

Extensive landslide and debris flow took a big toll on the landline network (Fig. 6.54 and 6.55). Severed optical fiber cables could be seen (Fig. 6.56) on the shoulders of the road leading to Beichuan. Some splice cases were just left on the side of the road (Fig. 6.57). Due to need to recover telecommunication quickly, installation of temporary optical fiber cables did not follow the normal practices required. Tight bend radii of the optical fiber cable was observed in many locations (Fig. 6.58 and 6.59). On the shoulder of the road close to Beichuan, spare utility poles pile could be seen (Fig. 6.60). Excess optical fiber cables were bundled on the pole (Fig. 6.61).

Fig. 6.60. On the right hand side of the road leading towards Beichuan, there were two piles of spare poles delivered for rebuilding.

Fig. 6.61 Optical fiber cables (in the circle) for emergency repair were bundling up on the pole.

6.5 Lessons Learned and Recommendations

The use of unreinforced masonry infill walls continues to cause damage to equipment inside switching offices in similar earthquakes throughout the world. In the Wenchuan earthquake, unanchored equipment and unrestrained circuit boards and switches caused extensive damage and service interruptions.

Providing engineered buildings and proper restraints for floor mounted equipment, push-in circuit boards, and switches with positive locking are some of the proven practices to reduce or to eliminate disruptions of telecommunication service.

The collateral damage to the landline system by collapsed buildings and bridges caused significant service interruptions. Although the service demand was reduced due to the destruction of these homes and businesses, the circuit interconnection capacity suffered.

The lack of backup power and the short duration of reserve power by batteries in remote cell sites resulted in many cell site outages.

In a province like Sichuan where majority of the areas are in mountainous terrain, telecommunication service providers should consider building small utility vehicles with cellular base station equipment powered by small generators for deploying to a disaster area to quickly restore service. Telecommunication is the primary lifeline that emergency services and rescue services use to coordinate their effort and resources allocation.

Educating the public about use of telecommunication service immediately following a disaster is extremely important. It results in having more circuits available to the emergency services to execute their work effectively. A control system can be established to manage calls in and out of the disaster area to reduce congestion and allow emergency calls to go through.

The service providers and the central government can establish a mutual agreement on repair priority, service priority, and equipment delivery priority from suppliers in case of emergencies such as damaging earthquakes or other natural disasters.

6.6 Acknowledgment

The authors appreciate the information provided by Zifa Wang of the Institute of Engineering Mechanics (IEM), Harbin, China, and permission to use photos in Figures 6.23 to 6.35, which were taken right after the earthquake.

All other photos used in this chapter were taken by the ASCE/TCLEE earthquake investigation teams (Phase I and Phase II).

6.7 Reference

Journal of Earthquake Engineering and Engineering Vibration, Vol 28 supplement, October 2008: General Introduction of Engineering Damage of Wenchuan Ms 8.0 Earthquake, Institute of Engineering Mechanics, China Earthquake Administration.

7 AIRPORTS

EXECUTIVE SUMMARY

There are five airports on the out skirts of Chengdu; two are for commercial airlines with schedule flights. The other airports are for military use.

Although Chengdu is about 90 km from the epicenter, the ground shaking level was quite low, about PGA = 0.05 g. Both commercial airports were shut down for a relatively short period and were reopened for scheduled flights. Many schedule flights were cancelled a couple of days later when relief materials had to be flown in for the victims of the earthquake. About six weeks after the earthquake, the airport returned to normal operations for passenger flights.

There was no electric power outage at these airports.

7.1 AIRPORTS

The international airport in Chengdu (Shuanliu (双 流)) (IATA: CTU, ICAO: ZUUU) suffered essentially no damage from the earthquake. Offsite power was lost to this airport almost immediately, but the on-site emergency generators worked, so there was no effect on airport operations. The level of ground shaking at this airport was likely low (perhaps 0.05 g).

It was reported that there was a 20 minutes shut down of the terminal due to loss of telecommunication (mobile phone network) and panic of the shaking. There was no damage to the terminal building and to the runway.

The airport was closed to commercial flights due to large number of relief supply flights traffic for two weeks a few days after the earthquake. By the end of May the airport was open to scheduled commercial traffic.

8 DAMS

EXECUTIVE SUMMARY

The M7.9 Wenchuan earthquake of May 12, 2008 was one of the most costly and damaging earthquakes in recent memory. The impact of this horrific event will be felt in the Sichuan region for generations and will require hundreds of billions of dollars for recovery, rebuilding, and restoration.

Vital to any community is a reliable and safe water supply; the Wenchuan earthquake damaged more than 2,600 dams and reservoirs of all sizes and types in eight Chinese provinces. Sichuan province experienced the majority of the damage to their dams and reservoirs with 1,996 dams—30 percent of the dams in Sichuan—experiencing some form of damage. Of the damaged dams, 69 were classified as Dangerous Situations of Dam-Break, 310 were classified as High Dangerous Situations, and 1,617 were classified as Secondary High Dangerous Situations. Damage to dams included cracks in 1,425 dams, collapse and settlement in 687 dams, slides in 354 dams, leaks and seepage in 428 dams, mechanical hoisting equipment damage in 161 dams, and outlet works, spillway, and office building damage to 422 dams.

In the modern era of dam design and construction, this extent of damage to a country's water supply and hydroelectric system has never before been realized or documented. This affords engineers a unique opportunity to evaluate the performance of these dams after subjection to significant earthquake loading and to learn from our findings.

This chapter looks at damages to dams in Sichuan region, including Zipingpu Dam near Duijanyan, Shapai Dam, and others damaged by the Wenchuan earthquake. It offers lessons learned for the design, construction, operation, and maintenance of these structures. This chapter will also compare Chinese dam design and safety practices with U.S. dam design and safety practices for earthen, concrete, and rockfill dams.

The May 12, 2008 Wenchuan earthquake was devastating. The 7.9 magnitude earthquake exacted a chilling toll on the people in the Sichuan province of China with 69,195 people confirmed dead, 18,392 people missing and presumed dead, 374,177 injured, US$86 billion in economic losses, more than 5 million left homeless, and 15 million evacuated. In total more than 45 million people were affected by this earthquake. The damage to the region's infrastructure is just as startling with more than 7 million buildings completely collapsed, 24 million buildings significantly damaged, water supply disruption to 10.6 million homes in 140 cities and towns, damage to more than 53,000 km of roads, and in excess of 48,000 km of water distribution pipelines and 2,666 dams and reservoirs damaged—1,996 of these in the Sichuan Province alone. This chapter discusses the damage to these dam reservoirs and, in particular, Zipingpu Dam, Shapai Dam, and Hongkou Dam.

The specifics of the Wenchuan earthquake have been well research and chronicled by others. What follows is only a brief summary of this earthquake. The Wenchuan earthquake struck the Sichuan Provence at 2:28 p.m. local time and registered 7.9 on the Richter scale. Shaking intensity reached its maximum in the Wenchuan area where a Modified Mercalli intensity of XI (Very Disastrous) was experienced. The intense shaking and the long duration of the earthquake almost completely destroyed the cities of Beichuan, Dujiangyan, Wuolong and Yingxiu. Figure 8.1 shows the pre- and post-earthquake photos of Beichuan City.

Fig. 8.1. Beichuan before (left) and after (right) May 2008 earthquake

The Wenchuan earthquake occurred on the Longmen Shan fault, which sits at the boundary between the Sichuan Basin to the east and the Tibetan Plateau and western Sichuan Mountains to the west (Fig. 8.2). In general terms, the regional seismicity of the area is a result of the India plate's northward convergence with the Eurasia plate at rate of approximately 50 mm per year. This convergence is also responsible for the uplift of the Himalayan Mountains southwest of Sichuan and the uplift of the Tibetan Plateau to the west.

This uplift created significant mountain ranges with large rivers cutting through them—the perfect scenario for building hydroelectric dams. The Sichuan province is the largest concentration hydroelectric dams in China and is one of its leading hydroelectric producers. The region also has thousands of irrigation and municipal drinking water reservoirs.

8.1 Overview of System Performance

There are 6,678 dams and reservoirs in Sichuan, and of these, 1,996 were damaged in the Wenchuan earthquake, which is roughly 30 percent of all dams in the province. The majority of damage was considered significant but not dangerous; however, the damage to 69 dams was considered dangerous with the risk of catastrophic failure, and the damage to 310 dams was considered to be highly dangerous. Table 8.1 shows the dams affected by the Wenchuan earthquake.

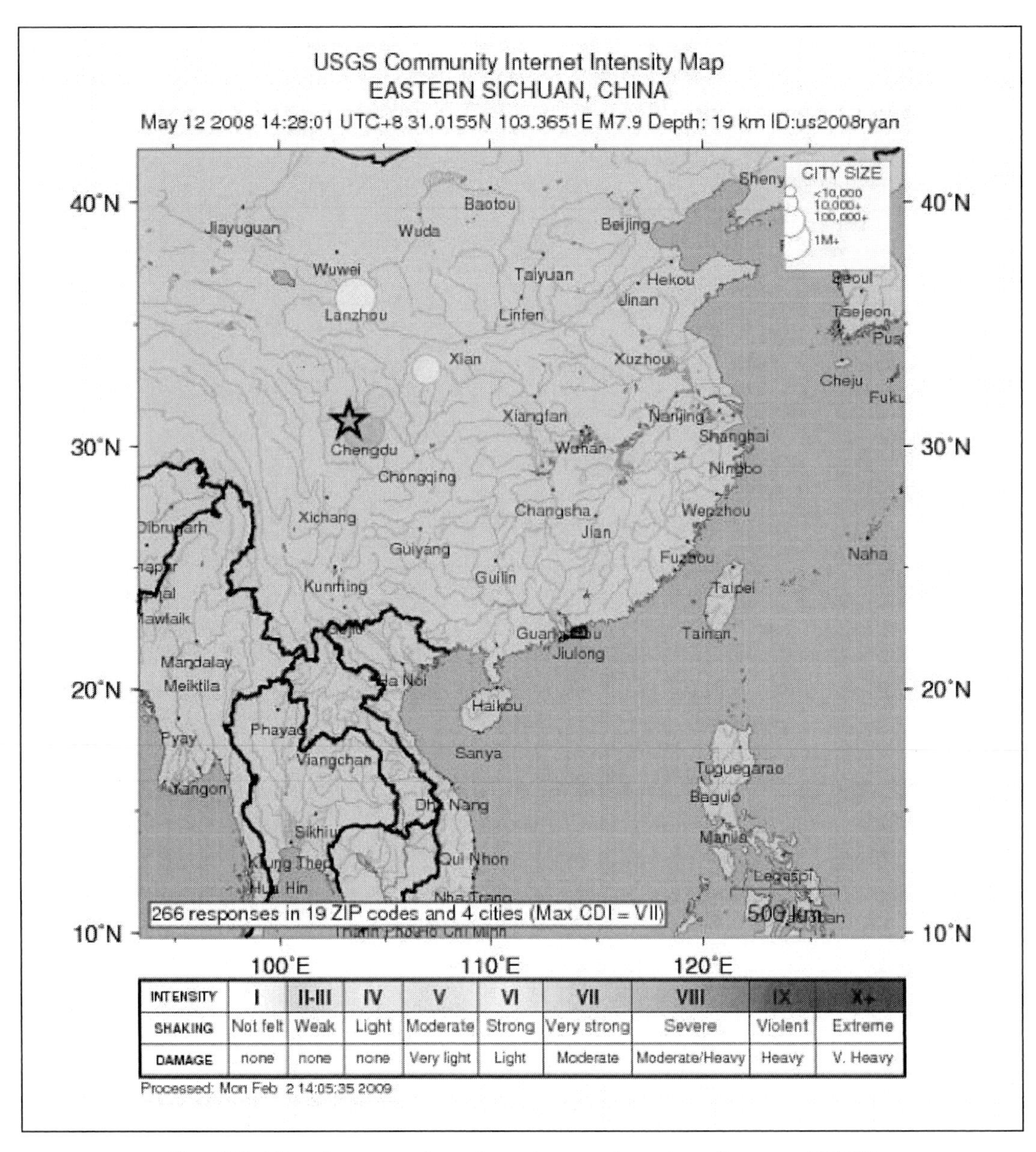

Fig. 8.2. Wenchuan earthquake epicenter location (Source: USGS)

This amount of damage to dams and reservoirs as the result of a single event is without precedence and a broad evaluation of these cases is beyond the reach of this report. Rather, this report will focus on several dams located near the observed surface faulting and where documented damaged occurred.

Table 8.1. Dams Affected by the Wenchuan earthquake

Order number	Name of province & municipality	Total number of reservoirs	Number of damaged reservoirs	Proportion of the damaged to the total for each province (city) (%)	Proportion of the damaged to the total in 8 provinces (municipalities) (%)	Classification of dangerous situations		
						Dangerous situations of dam-break	High dangerous situations	Secondary high dangerous situations
1	Sichuan	6678	1996	29.89	74.87	69	310	1617
2	Chongqing	2824	352	12.46	13.20	0	2	350
3	Shaanxi	1036	126	12.16	4.73	0	17	109
4	Yunnan	5422	51	0.94	1.91	0	2	49
5	Gansu	297	81	27.27	3.04	0	0	81
6	Guizhou	2105	12	0.57	0.45	0	0	12
7	Hubei	5804	25	0.43	0.94	0	0	25
8	Hunan	11435	23	0.20	0.86	0	0	23
Total		35601	2666	—	—	69	331	2266

Source: Journal of Earthquake Engineering and Engineering Vibration, *Vol. 28 supplement, Oct 2008: General Introduction of Engineering Damage of Wenchuan Ms 8.0 Earthquake, Institute of Engineering Mechanics, China Earthquake Administration.*

8.2 Performance of Sichuan Dams and Reservoirs

Zipingpu Dam

Perhaps the most heavily damaged dam in Sichuan was Zipingpu Dam, located on the Minjiang River directly upstream and 9 km from Dujiangyan City. Dujiangyan City has great historical significance to Sichuan Province as it is the site of one of the oldest manmade water control structures on earth. The 2,500-year-old Dujiangyan irrigation structure was designed and built to divert the majority of normal flows from the Minjiang River through developed areas of the city and surrounding farmland. During high-flow events, the structure altered the flow of the Minjiang River and diverted the majority of these flows away from the developed areas of Dujiangyan City, preventing damage to buildings and crops. Modern mechanical gates and control structures have since been added to the scheme, but the basic elements remain in service to this day.

The Zipingpu Hydroelectric Dam was constructed 9 km upstream of Dujiangyan City and is less than 20 km from the epicenter of the earthquake. Construction began in 2001 and was completed in 2006, including placing the hydroelectric generation online. The 156-m-high dam is a concrete faced rockfill dam (CFRD) and is used primarily for hydroelectric generation, although it provides flood control benefits as well. The reservoir created by the dam stores 1.112 billion m^3 (900,000 acre-feet) of water, and hydropower production is achieved using four turbines with a combined capacity of 760 megawatts.

Fig. 8.3. Zipingpu Dam; insert shows surface cracking on the down stream dam face.

At the time of the Wenchuan earthquake the reservoir contained roughly one-third of its normal capacity, estimated to be 0.38 billion m^3 (260,000 acre-feet). This low reservoir level allowed for the immediate inspection of a large portion of the structure, which would have been submerged had the reservoir been at full capacity. Measured ground accelerations at the reservoir were significant. The lower reservoir elevation also reduced the amount of seepage through the defects in the dam caused by the earthquake. These defects, and other damage to the dam, are described in greater detail below.

Shaking at the dam crest was intense and ground accelerations were measured to be 2.0 g, which indicates ground accelerations in the foundation ranged between 0.5 g to 1.0 g. This shaking caused maximum settlements at the dam crest of 73.5 cm (2.4 ft.), lateral displacement in the downstream direction of approximately 38 cm (1.25 ft.), significant cracking of the upstream concrete facing, cracking and displacement of the revetment block facing on the downstream face of the dam, and significant damage to the outlet works operator housing building. It was reported that damage to the outlet works prohibited the release of reservoir water until May 17 when repairs allowed for the release of water.

The original design of Zipingpu Dam anticipated maximum ground accelerations of 0.26 g, and at the crest, these estimates were exceeded by nearly an order of a magnitude. Nonetheless, the observed damages—especially the crest settlement and lateral displacements—were within tolerable limits. Properly designed and constructed rockfill dams are inherently robust structures and capable of remaining stable under extreme flood and earthquake loading conditions. When subjected to earthquake loading, embankment dams (including rockfill dams) typically behave by settling downward and displacing horizontally.

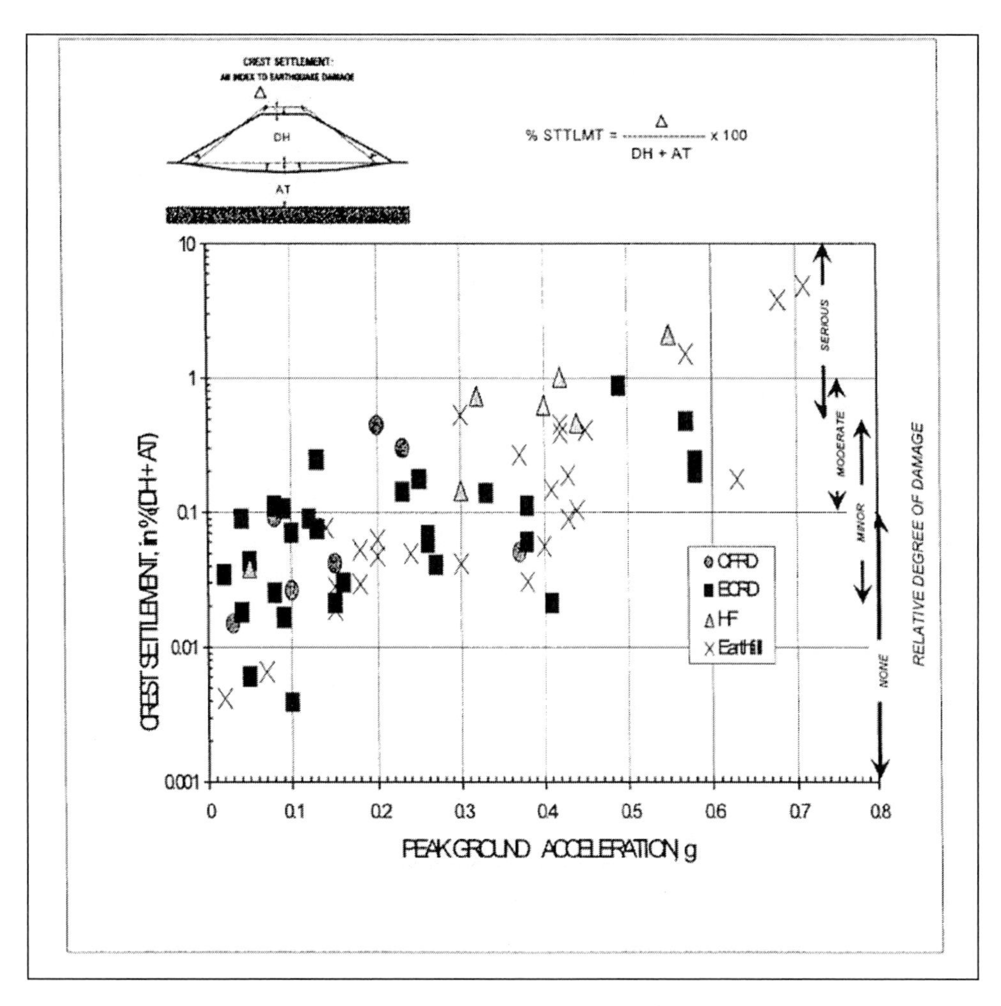

Fig. 8.4. Actual crest settlement vs. PGA (from Swaisgood)

Swaisgood (2003) developed an empirical methodology for estimating embankment dam deformation based on the historical performance of nearly 70 embankment dams subjected to earthquake loading. Using this empirical method and assuming an earthquake with a magnitude of 8.0 and peak ground accelerations between 0.5 g and 0.7 g, settlement would be expected to be on the order of 1 to 5 percent of an embankment dam's height.

At Zipingpu this translates to a predicted crest settlement, using the Swaisgood methodology of between 1.56 m and 7.8 m. Actual settlements were approximately 0.5 percent of the dam's height, less than half of the lowest predicted settlements, so it can be concluded that overall the Zipingpu Dam performed well given the violent shaking it experienced.

Figures 8.6 and 8.7 show other parts of the dam sustain damage.

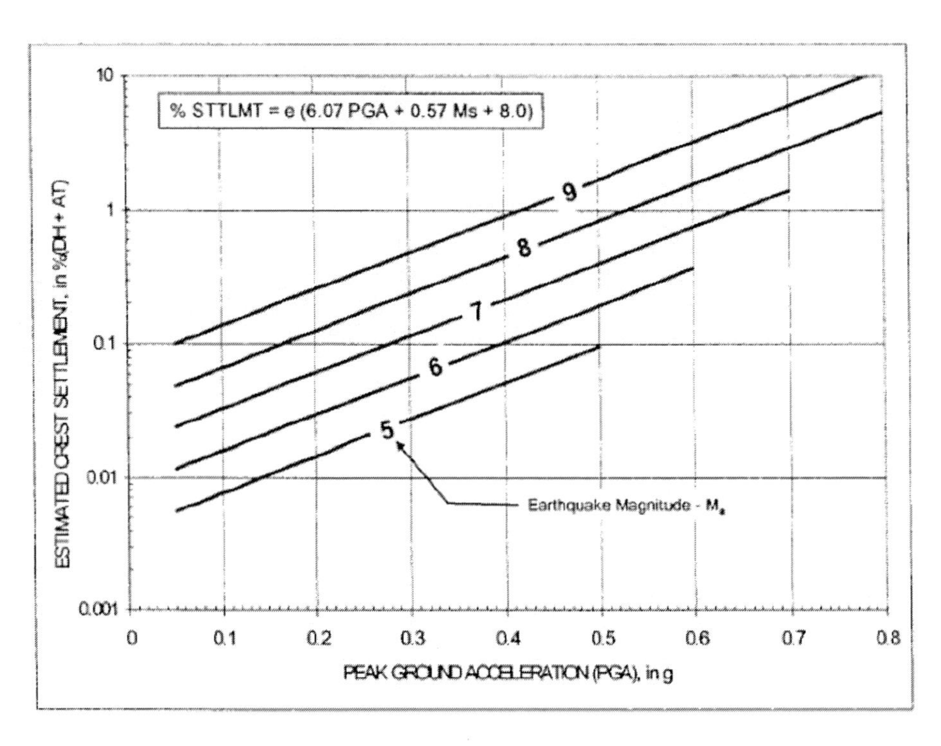

Fig. 8.5. Estimated crest settlement vs. PGA (from Swaisgood)

Fig. 8.6. Gate lifting equipment building sustained serious damage but did not collapse (Courtesy IEM)

Fig. 8.7. Collapsed guardrail on dam top (Courtesy IEM)

Shapai Dam

Shapai Hydroelectric Dam is a 132-m high roller compacted concrete (RCC) arch dam located upstream of Zipingpu Dam on the Chaopo He (Chaopo River) approximately 12 km from the epicenter of the earthquake. It is the world's tallest RCC arch dam. It was completed in 1999 and has a base thickness of 28 m or 21 percent of its height. Its reservoir holds 410 million m^3 of water, and it produces 36 megawatts of hydroelectric power and full capacity.

The most notable problem at Shapai Dam was simply gaining access to the structure. The remote location and the blockage of all roads to and from the site caused by landslide debris allowed for the initial inspection of the dam to be performed only via helicopter. When access was finally gained several weeks after the earthquake it was found to be in generally good condition despite the intense shaking it must have been subjected to. The dam was originally designed for a peak ground acceleration of 0.138 g, and while ground actual ground acceleration records are unavailable for the site, it is believed the dam experienced ground motions much higher than design levels.

Fig. 8.8. Shapai Dam (Courtesy of IEM)

The only significant damaged noted at the Shapai Dam site was limited to diversion pipes used to convey river flows around the dam. Falling boulders from the surrounding hillsides crushed these diversion pipes, which are no longer in use for river diversion.

As with properly designed and constructed concrete faced rockfill dams, roller compacted concrete dams that are similarly well designed and built are robust structures capable of safely withstanding extreme loading events.

Hongkou Dam

Hongkou Dam is located on the Yin River near the town of Hongkou, approximately 10 km north of Dujiangyan City and less than 20 km from the epicenter. Hongkou Dam was visually inspected by the team without the benefit of design drawings or construction records. The following describes the visual inspection.

Hongkou Dam was originally designed and built as a run of the river hydroelectric structure. It is not large, with an estimated structural height of roughly 10 m to 15 m. Reservoir releases are controlled by three low-level radial gates located near the left abutment and two upper-level radial gates located near the left abutment. An intake for hydroelectric production is located at the left abutment; however, hydroelectric generation no longer occurs at the dam because of the collapse of this intake tunnel, which occurred prior to the Wenchuan earthquake.

Fig. 8.9. Hongkou Dam after the Wenchuan earthquake

Minor cracking was observed at the main dam, but it was not clear if this cracking occurred before or after the Wenchuan earthquake. However, significant seepage is occurring around the right abutment through weep holes in a long retaining structure as well as through uncontrolled seepage zones further downstream. The total quantity of this seepage was estimated to be in excess of 0.03 m^3 per second (1 $ft.^3$ per second). It is believed this seepage predated the Wenchuan earthquake, but it is uncertain if seepage rates increased, decreased, or remained unchanged after the earthquake. No seepage measurement weirs, other monitoring, or instrumentation were noted during the inspection.

Outcrops of poor quality rock were noted on the right abutment, and it was surmised that seepage forces were flowing through this media with very little inhibition. While it is impossible to visually determine the stability of a dam, high rates, seepage, and their associated pressures are typically regarded as a destabilizing influence for most dams. High rates of seepage flows will eventually erode away the material they flow through. When this happens, catastrophic failure in the form of an uncontrolled reservoir release can occur more rapidly than repairs preventing it can be implemented.

Fig. 8.10. Hongkou Dam right abutment seepage

Earthen Dams

Numerous earthen dams were damaged as a result of the earthquake with the vast majority of these occurring longitudinally and parallel to the dam crest. In isolated cases transverse cracking was also noted but was not a predominate form of observed distress. Longitudinal cracking was observed mainly at the crest and on the upstream slope of the dam. Figures 8.11 to 8.16 show several earthen dams damaged in the earthquake.

8.3 China's Dam Safety Program

China has been building dams for more than 2,500 years and today has in excess of 85,000 dams and reservoirs in service—more than any country. China's dam safety regulation is organized under the responsibility of the Ministry of Water Resources. The country's dam regulation and safety laws were first organized and implemented in the early 1980's. These guides and standards included:

- general rules on reservoir management;
- reorganizing methods for measured data of earth dams; and
- guidelines for safety monitoring of concrete dams.

Fig. 8.11. Longitudinal cracks at crest of newly constructed earthen dam, Fenggu Town, Mianyang City (Courtesy IEM)

Fig. 8.12 Longitudinal cracks at crest of earthen dam, Changdaogou Reservoir, Mianyang City (Courtesy IEM)

Fig. 8.13. Longitudinal cracks at crest of earthen dam, Dasongshu Reservoir (Courtesy IEM)

Fig. 8.14. Upstream slope movement of earthen dam, Bailin Reservoir, Hanwang Town, Mianzhu City (Courtesy IEM)

Fig. 8.15 Longitudinal cracks at crest of earthen dam, Fengshou Reservoir, Anxian County (Courtesy IEM)

Fig. 8.16. Upstream slope movement of earthen dam, Fengshou Reservoir, Anxian County (Courtesy IEM)

1988 saw the first water laws enacted in China, and from these laws the first dam safety programs were established. The first laws and dam safety guidelines were adopted in the early 1990s. Since then the following standards have been used for regulating dam safety in China:

- dam safety appraise methods;
- dam and reservoir registration methods;
- guidelines for dam safety assessment;
- regulation on maintenance and mending of earth dams;
- regulation on maintenance and mending of concrete dams;
- general rules on reservoir comprehensive operation;
- evaluation rules on reservoir operation in floods;
- methods of reservoir downgrade and retirement;
- regulation on examination of hydraulic project management; and
- standards of examination on reservoir management.

The understanding and evaluation of the risks associated with dam and reservoir operation is also becoming more prominent in China. As the population and economy continues to grow and expand, the risks associated with dam and reservoir operation during both normal and unusual conditions becomes more critical to understand. Regulations are currently being developed to assess and understand these risks.

When compared to the dam safety rules, regulations, and regulating agencies of other countries, the Chinese dam safety program is relatively young. However, the Chinese program appears to be improving. Chinese engineers are regular participants in world dam safety conferences, regularly write and contribute scholarly papers on dam safety and dam engineering, and have established internal organizations, such as the Chinese National Committee on Large Dams (CHINCOLD), that promote the safe design, construction, operation, and maintenance of dams throughout China and the world.

8.4 Conclusions and Recommendations

The Wenchuan earthquake has provided engineers who work on dam design and dam safety issues invaluable insight into the performance of both high- and low-head, earthen and concrete dams to date. No other earthquake event affected this large number of dams. While some dams were severely damaged, no uncontrolled reservoir releases occurred, events that likely would have killed hundreds of thousands of people with little or no warning.

There is little doubt that China's dam safety laws have improved the safety of these structures and the continued refinement and improvement of these laws and regulations will only serve to increase the safety of China's dams. To this end, the author recommends two specific areas of future attention.

The first recommendation relates to the level of peak ground accelerations that China's dams are designed to resist. The Zipingpu and Shapai Dams were each subjected to peak ground accelerations that far exceeded their design assumptions. In high-seismic zones, international practice often requires dams be designed to resist peak ground acceleration in excess of 1.0 g. Zipingpu Dam, which is obviously in a very high seismic zone, was designed with an assumed peak ground acceleration of

0.26 g. Similarly, Shapai Dam was designed with an assumed peak ground acceleration of 0.136 g. A more conservative approach to peak ground accelerations will provide additional safety to China's dams.

The second recommendation relates to the development of more definitive emergency action plans. Dam safety emergencies are terrifying events, and having a plan in place to follow when most people involved are panicked has been shown to be an extremely valuable tool. Emergency action plans dictate specific actions, notifications, downstream populations at risk, and when necessary, safe evacuation routes for those in danger. These plans also increase the speed and efficiency of communication, which is a vital element in helping reduce potential casualties in the event of an uncontrolled reservoir release.

8.5 Acknowledgment

The authors appreciate the information provided by Zifa Wang of IEM (Institute of Engineering Mechanics), Harbin, China, and permission to use the photos presented in Figures 8.6 to 8.7 and Figures 8.11 to 8.16.

All other photos used in this chapter were taken by the ASCE/TCLEE earthquake investigation teams (Phase I and Phase II).

8.6 Reference

Journal of Earthquake Engineering and Engineering Vibration, Vol. 28 supplement, Oct 2008: General Introduction of Engineering Damage of Wenchuan Ms 8.0 Earthquake, Institute of Engineering Mechanics, China Earthquake Administration.

9 SCHOOLS AND GENERAL BUILDING STOCK

EXECUTIVE SUMMARY

The area is mountainous with peak elevations greater than 4,000 m (approximately 12,000 ft) with steep slopes. Numerous slides extend up the slopes hundreds of meters and contained massive boulders that contributed to the down slope damage to roads and lifelines adjacent to the roadbed in many locations. In a few locations, landslide debris covered the roads to depths exceeding 10 m. At Beichuan, a combination of rock-fall and deep-seated slides and debris flows destroyed perhaps 20 percent of the town (population estimated 40,000 people), and strong inertial shaking damaged much of the rest of the town. As of October 19, 2008, the town remained in its post-earthquake condition with no attempt to excavate the landslide/debris flow debris or extricate the dead. Reportedly, the town of Wenchuan was in similar condition, as was at least one other village. Slope failures continued to occur during the recovery and reconstruction operations in unstable areas, with additional landslides being triggered by ongoing seasonal rains.

The primary damage occurred in Sichuan Province, whose capital is Chengdu. The province's population was 83.29 million, according to the 2000 Census. The population density is higher in the east, and much lower in the west.

Neighboring provinces are Hubei, Guizhou, Yunnan, Xizang, Qinghai, Gansu, and Shaanxi. Several of these provinces sustained losses of varying degree of severity; Gansu suffered the most losses.

To appreciate the issues related to lifelines damage and their restoration in this earthquake, we must also appreciate the damage to the general building stock. The restoration times for lifelines in this earthquake are extraordinarily long. For example, electric power was not restored in some areas (including heavily industrialized areas) with the strongest shaking (PGA >0.3 g) for more than 160 days after the earthquake. This was in large part due to the vast destruction to the general building stock, leading to total collapse of the local economy.

Perhaps 70 to 80 percent of all damage was due to inertia loading. Less than 0.5 percent was due to fault offset, and the remainder was due to landslide. Much of the inertia-loading damage can be attributed to the essentially complete lack of seismic load design in single family residential or multi-family residential buildings and the grossly inadequate seismic design for engineered structures, primarily highway bridges, tunnels, and electric substations. Most of the landslide-induced damage within towns, which may have a population up to 100,000 or more, could have been avoided by zoning regulation as the locations with massive landslide-induced damage were readily observed to be in high-risk landslide/debris flow areas. On several occasions, the ASCE team inquired about the seismic loading provisions of local building regulations. In nearly every case, we were told that because the affected area had no reported earthquakes within the last 300 years, the seismic design requirement for the area was a Level 5 on a seismic design scale of 1 to 11. At Level 5, the seismic design requirement would be equivalent to about PGA = 0.05 g to 0.10 g. This gross under-design resulted in extremely high damage rates to engineered structures (elevated bridges and larger multi-story residential structures) wherever PGA was greater than 0.4 g (estimated at 35 percent of structures having reached the complete damage state). Of the remaining 65 percent of engineered structures, essentially all had some form of damage, ranging from slight (15 percent) to moderate (20 percent) and major (20 percent). Note, all percentages are estimated based on limited field observation; damage states correspond to those described in HAZUS99, the FEMA the natural hazard loss estimation methodology software program. This is not to say that there were no engineered structure successes. We observed (and drove over) many single span highway bridges that suffered no more than minor damage (approach settlement, minor abutment damage, and concrete railing damage).

For non-engineered structures, the damage rates were similar to those for engineered structures. These non-engineered structures are single-family residences, which are commonly built by the building resident. In the rural mountainous areas, which had the highest ground shaking, there are many small villages (population of 500 to a few thousand or so). In these villages, the common style of residential construction is one-floor (sometimes two-floor) rectangular buildings built of unreinforced masonry (brick) walls with 0.33-in. by 3-in. wood roof trusses covered by roof tiles. The roof tiles rest on the wood slats by gravity only (no positive attachment). These types of buildings faired very poorly, with collapse rates of 50 percent or higher in the strongest shaking areas. A smaller percentage (perhaps 10 percent) of houses in villages are built with traditional wood construction, consisting of vertical wood poles lightly nailed to horizontal wood poles, with roof systems similar to those of brick houses, and walls made from lightweight materials.

Three provinces were affected (Sichuan, Gansu, and Shaanxi). More than 40,000,000 people were affected in some manner by the earthquake. As of September 2008, it is estimated that nearly 89,000 people were killed and 370,000 injured by the earthquake. Chinese authorities estimate the economic impact at more than 700 billion RMB (approximately US$100 billion)—an enormous impact to the local region.

The CEA reports the damage statistics in Table 9.1 for buildings in areas with different MMI zones. Table 9.2 provides estimated economic impacts.

Table 9.1. Damage to Building Stock

MMI Region	Structure Type	Collapse	Heavy	Medium	Slight	No Damage
IX+	Frame	28.23	22.86	21.01	14.84	13.06
	Brick + RC	29.32	18.77	44.95	6.49	0.47
VIII	Frame	3.8	19.35	24.7	20.78	31.37
	Brick + RC	10.81	21.89	23.73	19.26	24.31
VII	Frame	0.3	1.75	5.99	9.64	92.32
	Brick + RC	2.95	5.37	10.58	26.4	54.7
VI	Frame	0	0.3	2.13	5.27	92.3
	Brick + RC	0.91	1.94	4.32	10.28	82.55

Table 9.2 shows the damage statistics for Dujiangyan, the largest population center near the epicentral region. The survey was prepared by randomly sampling about 1,000 structures in various locations around the city. Damage was recorded by visual external survey. Three styles of building construction were used, namely "small house," "frame structure," and "frame VII." The following describes these three categories:

- Small houses are generally one or sometimes two stories, are made of brick walls and a timber (poles) roof, and are covered with tiles. They are non-engineered. Over larger spans, timber roof trusses might be used. In a few cases, the entire house is made from vertical wooden poles with perhaps metal sheathing. We did not observe use of plywood or wood stud walls, so that timber structures in the area should not be considered comparable to wood houses common in the United States. We saw many of these buildings being reconstructed after the earthquake using the same style of construction as used pre-earthquake.

- Smaller frame structures include reinforced concrete columns and beams, infilled by unreinforced brick. At lower floor levels, one or more walls might be open (glass walls, store fronts). These are non-engineered.

- Larger frame structures are intended to reflect engineered structures to the local seismic code (Intensity VII, PGA = 0.1 g). These might be multi-storied, monumental, government-owned structures. Mostly, they are built using the same methods as smaller frame structures but possibly with better detailing and avoiding soft first stories. We saw no evidence of ductile detailing such as closely spaced stirrups.

As evidenced in Table 9.2, the frame VII structures generally performed better than the non-engineered structures. Whether this is entirely due to the style of construction is unclear, as most large engineered structures are located in the eastern portions of the city and thus were likely exposed to somewhat lower levels of shaking.

Table 9.2. Damage to Building Stock, Dujiangyan (PGA = 0.2 g-0.3 g)

Structure Type	Number	No Damage	Slight	Medium	Severe	Collapse
Small House	230	24%	7%	4%	30%	35%
Frame	224	18%	8%	23%	46%	5%
Frame VII	551	46%	21%	15%	17%	1%
Total	1,005					

Table 9.3 shows the performance of structures in the largest nearby city, Chengdu, which has a population of many millions. Ground motions were about PGA = 0.08 g (western portion) to 0.06 g (eastern portion). When observed from the interior, our informal survey of perhaps 10 buildings indicated small fresh cracks at corners of buildings in the northwest quadrant of the city in about 10 percent of the cases, so the results in Table 9.3 should be indicative only of rapid external observations for the "slight" and perhaps "medium" categories. While the survey was performed by local engineering companies, we are unsure of how a 0.03 percent value can be established given a sample size of 400 structures as 1 in 400 is 0.25 percent. Thus, the reader is cautioned to evaluate the data with care.

Table 9.3. Damage to Building Stock, Chengdu (PGA = 0.06 g-0.08 g)

Structure Type	Number	No Damage	Slight	Medium	Severe	Collapse
Small House		97%	3%	0.04%	0.01%	0.03%
Frame		97%	3%	0%	0%	0%
Frame VII		99%	1%	0%	0%	0%
Total	400					

Table 9.4 shows similar results for Mianyang (Fucheng and Youxian areas). These areas are thought to have experienced round motions with PGA = 0.1 to 0.15 g or so.

Table 9.4. Damage to Building Stock, Mianyang (PGA = 0.1 g-0.2 g)

Structure Type	Number	No Damage	Slight	Medium	Severe	Collapse
Small House		56%	19%	15%	8.4%	1.9%
Frame		69%	18%	13%	0.7%	0%
Total	199					

Note: See map, Fig. 2.22.

Table 9.5 shows similar results for Deyang. These areas are thought to have experienced round motions with PGA = 0.1 to 0.15 g or so.

Table 9.5. Damage to Building Stock, Deyang (PGA = 0.1 g-0.15 g)

Structure Type	Number	No Damage	Slight	Medium	Severe	Collapse
Small House		39%	29%	6.8%	22.2%	3.4%
Frame		50%	39%	8.0%	2.4%	0.1%
Frame VII		77.5%	19%	2.9%	3.3%	0%
Factory		72%	26%	0.9%	1.7%	0%
Total	199					

Note: See map, Fig. 2.22.

Table 9.6. Economic Impacts (Chinese data)

Province	Earthquake Loss (RMB, Billion)	2007 GDP (RMB, Billion)	Loss / GDP (%)	Loss / China GDP (%)
Sichuan	755.00	10,505.3	58.8	2.5111
Gansu	55.28	2,699.2	16.4	0.1800
Shaanxi	24.814	5,369.9	4.25	0.0927
Chongqing	5.423	4,111.82	1.32	0.022
Yunnan	1.682	4,721.8	0.36	0.0068
Ningxia	0.083	834.16	0.10	0.0003
China	845.1	24,600.00		2.81

Chinese authorities have suggested the direct costs to rebuild as listed in Table 9.6. For conversion, 6.8 RMB = US$1, as of October 2008. We suspect that all values are inflated by 10 times as the China total of 246,000 billion RMB would be about US$36 trillion, which is about 10 times higher than reality. While we have not made the correction, the loss rations (2.81 percent of China gross domestic product) would factor out this error.

The ASCE/TCLEE team did not observe loss ratios as high as reported in Table 9.6. We observed nearly 100 percent direct damage rates in the mountainous towns and villages, but these areas represent perhaps 2 percent of the population of Sichuan province (population nearly 43,000,000 people). For example by October 2008, the major city of Chengdu was essentially at 100 percent of its pre-earthquake economic activity. Therefore, the Earthquake Loss column should be viewed cautiously, and the total loss versus China GDP of 2.81 percent should be considered a long-term loss to rebuild the affected area to its pre-earthquake condition, which might take a decade or longer in devastated towns such as Beichuan.

In the following sections, we present just a few selected photos of building damage observed by the team members. This selection was developed to highlight key issues.

9.1 Schools

The collapse of schools and fatalities of school children in this earthquake was particularly devastating. The number of fatalities varies from 7,000 to 9,000. Regardless of the actual number, one fatality is one too many. The focus must be on improving both the zoning requirements and the building code to ensure life safety is on the top of the criteria for school buildings.

From talking to various schoolmasters and teachers during our visit, it became obvious that training for both teachers and students to act effectively and efficiently in an earthquake is a must. One school in Nanquan (南泉） where no student was inured resulted from the efficient evacuation of students. Part of the school building collapsed soon after the students were outside.

Yingxiu (映 秀)

Figure 9.1 shows the collapse of main building at the entrance of Xuan Kou (漩口） High School in Yingxiu. This school has many buildings on its campus, and except for the student residence, most had collapsed or partially collapsed. The damage to school buildings, however, was not obviously better or worse than collapses to the general building stock nearby.

This school is a relatively new building. The most important cause of failure is the seismic zoning criteria; it is low compared to the history of strong ground shaking in this province. In addition to this low requirement, the absence of importance factors in the building code of critical structures is also key. Without the original structural building plan, it was impossible to evaluate quality and workmanship of the construction. Based on the ruins, the quality of construction was not substandard. The failed building figures shown in this section reveal the inadequate design of basic load transfer elements and load bearing capacity. This is a hard lesson and will be remembered for years.

Fig. 9.1. Partially collapsed Yingxiu, Xuan Kou High School.

Fig. 9.2. Totally collapsed building Yingxiu, Xuan Kou High School.

Fig. 9.3. Yingxiu, Xuan Kou High School student residence was heavily damaged but did not collapse.

Fig. 9.4. Other building within the school campus sustained extensive structural damage. Many students were injured and killed.

Fig. 9.5. A detailed view of the reinforcement connection to beams and joist.

Nanquan (南泉)

This is one of the four schools that we visited with a happy story. Due to the quick thinking and fast reaction of the teachers and staff of the school, the students were quickly evacuated from the classrooms. The result is that there was no fatality or injury. Figures 9.6 to 9.10 show the remains of the school and the temporary school that was built quickly so that students could go back to school.

This school is another unreinforced masonry (URM) three-story building. It was impossible to identify its main structural elements from the ruins. We can only conclude from failures of similar structures that this building was not designed and constructed with seismic load bearing capacity. Simply, URM does not have lateral load capacity to resist earthquake force.

Sadly, the Nanquan schoolmaster's daughter, a teacher at one of the schools that collapsed, was killed while trying to protect students from falling debris.

Fig. 9.6. The front gate of the Nanquan primary school. The sign shows red brick for sale at 0.20¥ per brick.

Fig. 9.7. Piles of bricks are the only remains of this primary school. Only a classroom desk indicates that this was a school. The one-story building used for student activities in the background did not collapsed; however, there were cracks on outside walls.

Fig. 9.8. This figure shows the concrete floor slab used. Note the small size reinforcement bar.

Fig. 9.9. Tents were erected in the school basketball field to store undamaged desks, chairs, and other materials used in classrooms.

Fig. 9.10. New buildings constructed as the temporary classrooms across the rice field. The banners (insert) express thanks and celebrate the smooth construction of these building.

Shifang (什邡), Longju Zhen (龙 居 镇)

The primary and high school in Longju Zhen sustained extensive damage. The three-story classroom building collapsed while the office building on the left-hand side of the school entrance sustained structural damage but remained standing (Fig. 9.11 to 9.12). The local police did not allow us to enter the campus due to the risk from poor hygienic conditions after the bodies were removed.

A temporary shrine was set up at the front gate by the parents whose children lost their lives (Fig. 9.13). Most of the students were 9 to 11 years old.

The inspection report on the causes of collapse was also posted on a board in the temporary shrine. The report details failed elements, such as sizes and materials. Based on the report it was not difficult to conclude that the failed structures lacked seismic design details.

Fig. 9.11. Shifang, Longju Zhen primary school campus. On the left-hand side, covered with a plastic sheet with stripes, is the temporary shrine in memory of the students killed on this campus.

Fig. 9.12. The collapsed three-story classroom building.

Fig. 9.13. Temporary shrine set up by the parents of students who were killed by the collapse of the school building.

Beichuan (北 川)

The school we visited in Beichuan was located about 5 km from the city. The ruins were fenced in (Fig. 9.14), as the whole building had collapsed. There was no information about casualties at this school, which had been relocated close by.

Another school in Beichuan town had been completely demolished by rock fall, leaving only a basketball goal as a sign that a school had existed (Fig. 9.15). This was the worst of all the schools affected by the earthquake. Almost all of its students died.

Fig. 9.14. This school is located on the main road leading to Beichuan town.

Fig. 9.15. The remains of this a school in Beichuan town are under the rocks. Note the basketball goal on the left hand side.

9.2 Beichuan (北 川) Town

Photos from Chinese newspapers show people wandering about in central Beichuan one minute after the earthquake. Great amount of dust in the air due to the landslides lingered for a long time as the town is situated in a valley. Members of ASCE/TCLEE Phase II team visited the site on October 18, 2008, but the PLA did not allowed us to enter because of the great number of bodies still buried in the rubble. After the May 2009 Chengdu lifeline loss reduction symposium, the ASCE presentation team were allowed to enter the town and observed the damage that was not cleared. The town will be closed forever and will be set up as memory site of the earthquake. A burial site was set up as the monument to pay respect to the victims buried by the collapsed buildings, massive landslide, and rock falls caused by the earthquake.

Figures 9.16 and 9.17 show the town of Beichuan before the earthquake. See Figure 2.30 in Chapter 2 for an aerial view of the landslide zones. Figures 9.16 and 9.17 were taken from a slope overlooking the town, with the vantage point at the bottom center of Figure 2.30 looking northeastwards. In the foreground in Figure 9.17, there are three- and four-story buildings covered almost to their roof tops by a large debris flow. On the right, a large landslide/rock fall has impacted several multi-story buildings (including a school) leading to their collapse. There are numerous tilted buildings around the town, likely due to inertial overload, although in this town, the potential for liquefaction-induced foundation failures might be plausible. Some building collapses are completely covered by the debris flow. Figure 9.18 shows an uphill view of the primary source for the debris flow.

The ASCE/TCLEE team members who attended the May 2009 lifeline loss reduction symposium in Chengdu visited Beichuan with special permission obtained by the Sichuan Association of Science & Technology in Mianyang. The entire town is designated as a memorial site of the earthquake. About 85 percent to 90 percent of the building stock in the city sustained damage of various degrees. We estimated that about 30 percent of the damaged buildings were caused by landslide, rock fall, and debris flow. See Figures 9.19 to 9.26 for typical damage. Figures 9.27 to 9.30 show damaged buildings as a result of strong shaking.

One apartment building had smoke marks around one of its windows indicating that a fire had broken out (Fig. 9.31).

Fig. 9.16. Beichuan before the earthquake

Fig. 9.17. Beichuan 160 days after the earthquake

Fig. 9.18. Beichuan, upstream source for debris flow

Fig. 9.19. The buildings along this foothill are under the pile of rock that fell from the top of the hill.

Fig. 9.20. View from a different angel to show the extent of rock fall and the debris of buildings being pushed to the low land

Fig. 9.21 Landslide that damaged and buried a number of buildings along its path.

Fig. 9.22. Large part of the town buried by debris flow

Fig. 9.23. This photo shows the force of the debris flow. The splash shot up two floors.

Fig. 9.24. The debris flow carried rocks that did the most damage.

Fig. 9.25. Timber debris flowing from the mountains also caused lots of damage to buildings.

Fig. 9.26. Large boulders from the mountains were also a major cause of building damage.

Fig. 9.27. Not all buildings were damaged; here one remained standing while the one next to it collapsed.

Fig. 9.28. Buildings damaged due to strong shaking

Fig. 9.29. This building pancaked with four floors collapsing to the second floor.

Fig. 9.30. This partially finished building leaned backward along with major structural set back. It was empty with not load on it.

Fig. 9.31. Soot marks the location of a fire, which most likely started on the third floor.

9.3 Hongkou (虹口)Area

This small township is very close to the epicenter. Surface faults were observed in the area. A large number of URM structures were damaged, either totally collapsing or suffering cracks on all walls. The area experienced major landslides and rock falls, blocking the only road access to the remote villages in this township.

Figures 9.32 to 9.37 show the various conditions of buildings along the major road leading to Hongkou. Almost all of them were URM or concrete frame with infill wall structures. A row of buildings experienced minor cracks on walls, one of which was a restaurant that was open for business during the ASCE/TCLEE Phase I team's visit (Fig. 9.38 to 9.40).

The military was setting up relief areas to provide the local population with sanitary facilities and water. They were also constructing temporary shelters for victims who lost their houses.

Many houses and roads in a small village located a few meters from the surface fault suffered extensive damage from both strong ground shaking and falling rocks (Fig. 9.41 to 9.47). Boulders broke through the back wall of a rest stop and, in one case, rolled through the front wall of a hotel in Hong Kou Village (Fig. 9.45 and 9.46). The hotel workers were using portions of this building, as of October 20, 2008. The rock pile in Figure 9.46 was created by breaking the original boulder into small pieces, either to off haul or for re-use as building materials. The area had a surface fault about 2 km long and suffered a few minor landslides.

Fig. 9.32. Remains of a rest stop shelter on the road to Hongkou that was demolished by rock fall.

Fig. 9.33. Professor Manchao He and one of the boulders that came through the rest stop.

Fig. 9.34. Collapsed timber roofs of the URM houses. The tent was set up by the PLA for the victims and those who did not want to return to their houses due to large number of strong aftershocks.

Fig. 9.35. Houses reduced to rubble close to Hongkou town. There were many casualties in this area.

Fig. 9.36. A strip of houses that sustained rather minor damage. Note tents were set up for people whose houses were destroyed.

Fig. 9.37. These houses showed significant structural damage but did not collapse.

Fig. 9.38 The investigation team taking a short break at this temporarily rebuilt restaurant. Note the pile of rubbles from a collapsed building on the left just behind Jian Wang, a Ph.D. student and our local support from SW Jiaotong University, Chengdu.

Fig. 9.39. This restaurant in Hongkou, where the team stopped for lunch, performed quite well with very minor superficial damage.

Fig. 9.40 Team members stop for lunch in Hongkou. Note the inside of this restaurant did not show any sign of damage. Some minor cracks were noted.

Fig. 9.41 One building in a group of resort-style hotel buildings on a hill side above a creek. See Fig. 9.43 for the front of the building.

Fig. 9.42 The front of the building in Fig. 9.42 shows significant structural damage; however, it did not collapse.

Fig. 9.43. The courtyard the hotel was littered with boulders that rolled down from the mountain above. The house on the right was damaged by the boulders.

Fig. 9.44. This new building on the north side of the court yard showed some minor cracks. The second floor was not yet occupied, and the railing had not been installed.

Fig. 9.45. Two boulders entered the hotel dining room through the wall causing lots of damage. Luckily the boulders missed the columns; otherwise, the building would have collapsed.

Fig. 9.46. One boulder broke through the back and front wall.
It was cut into smaller pieces before the team's visit.

Fig. 9.47. Another boulder impacted one of the building's exterior brick walls.

9.4 Dujiangyan

Dujiangyan is about 60 km from Chengdu, the capital of Sichuan Province, and about 30 km from the closest fault rupture. It is one of the major cities in the earthquake area that sustained heavy losses in both property damage and human lives. Significant damage to water system, electric power system, and telecommunication system were reported. More than 20 percent of the modern buildings sustained damage from stress cracks to collapse. Most of the URM buildings did not survive the earthquake.

Figures 9.48 through 9.61, taken by the two ASCE/TCLEE team members, present a wide range of building in the city from those that suffered major damage to those that performed well.

Because of the large number of strong aftershocks (between M_s = 5 to 6), many residents whose buildings were undamaged did not want to return to their apartments and stayed in tents set up in open spaces in the city instead.

Fig. 9.48. Two-story URM building with pre-cast concrete floor slabs collapsed, which was common in the older part of the city.

Fig. 9.49. Collapsed brick infill walls, another common scene within Dujiangyan.

Fig. 9.50. This concrete frame building performed well except for the masonry railing on the third floor.

Fig. 9.51. The third floor of this building fell about 1 m to the second floor and shifted to the right.

Fig. 9.52. This concrete building performed quite well; only minor stress cracks were observed on the outside. The windows were not broken.

Fig. 9.53. This concrete frame building also performed well. There were stress cracks, and the plaster wall fell off the outside of the building.

Fig. 9.54. A two-story URM building collapsed. The building behind and next to it sustained minor damage. A window was broken on the top floor the building in the background.

Fig. 9.55. A three-story restaurant building collapsed. Part of the tiled roof shows on the right.

Fig. 9.56. More than one-third of this five-story apartment building collapsed. The remaining structure was condemned as unsafe.

Fig. 9.57. The second floor of the building collapsed with the top four floors resting on it. It appears that the remaining structure is leaning towards the collapsed section.

Fig. 9.58. The second and third floor of this apartment building is a shopping mall with a high ceiling, creating a soft-floor effect. Most of the damage is on these two floors. Note the windows above are not damaged.

Fig. 9.59. This Li Bing Temple had significant damage but did not collapse.

Fig. 9.60. Undamaged gate in Dujiangyan. There were minor cracks on the exterior wall of the building on the left.

Fig. 9.61. Another old style gate was not damaged. Note the building on the left.

9.5 Yingxiu

Yingxiu is a small town about 12 km due west of Dujiangyan. By road it is about 20 km away. It is very close to the epicenter and the fault.

Close to 95 percent of the buildings in this town were destroyed. Access to the town was cut off due to collapsed bridges, tunnels, and landslide covering roads. Electric power was cut off due to the total collapse of Ertaishan substation. Telecommunications were out of service due to power outages, severed connection cables, and damaged cell antenna.

Many buildings in the Xuan Kou High School campus collapsed (see section on Yingxiu above).

The following figures were taken by members of ASCE/TCLEE Phase I, Phase II, and the May 2009 Chengdu lifeline loss reduction symposium team.

Fig. 9.62. Partially and totally collapsed buildings were seen everywhere in this small town.

Fig. 9.63. A collapsed manufacturing facility with a standing, but heavily damaged, office building in the background.

Fig. 9.64. Most of the buildings in Yingxiu were reduced to rubbles.

Fig. 9.65. Devastation of another part of Yingxiu

Fig. 9.66 Another view of the town

Fig. 9.67. A cement manufacturing facility close to Yingxiu showed very little damage. It was not operating during the team's visit due to road damage.

9.6 Mianyang

The ASCE/TCLEE Phase I team stayed in Mianyang one night on way to Nanba town. Mianyang is about half way between Chengdu and Nanba.

The team stayed in a hotel close to the city center. While we did not have time to investigate, we were able to check some damage to some buildings from the outside. The hotel experienced strong shaking, and cracks in the rooms were evidence of the damaging earthquake. The building adjacent to this hotel sustained extensive damage.

Team members felt an aftershock ($M_s = 4.3$) during the night we stayed in Mianyang.

Fig. 9.68. Cracks on walls inside the hotel rooms were evident of strong shaking in this city.

Fig. 9.69. The building next to the hotel ASCE/TCLEE Phase I team members stayed was damaged and closed.

9.7 Chengdu

Chengdu did not experience any strong earthquake motion. Very minor damage was noted, but there were no collapse buildings. The figures in this section show very minor damage from the earthquake.

Fig. 9.70. Tiles fell of the roofs of these buildings in the center of Chengdu.

Fig. 9.71. Glass windows from this building fell on the roof of the entrance.

Fig. 9.72. A few large advertisement signs fell on the roof of this building in Chengdu.

9.8 Shifang

Shifang is a manufacturing county in Sichuan. Most of the manufacturing facilities are chemical plants producing fertilizers, cement, and industrial chemicals.

The manufacturing facilities suffered extensive damage. During our trip, we saw only one cement manufacturing plant in operation. It showed no signs of any damage to the silos or buildings.

Most of the residential buildings in Shifang, which are small URM-type buildings, were severely damaged or collapsed.

Fig. 9.73. The raw material silo at this chemical processing plant in Shifang collapsed at about mid-height. This is the only silo in this plant that was damaged.

Fig. 9.74. The silos for processed material, located about 200 m apart, showed no signs of damage. The damaged silo is located to the right of this group of silos.

Fig. 9.75. A fertilizer plant in Shifang sustained extensive damage to most of its buildings.

Fig. 9.76. Damage at another building of the fertilizer plant shown in Fig. 9.76.

Fig. 9.77. Extensive damage at another chemical processing plant in Shifang.

Fig. 9.78. The building of this chemical plant was leaning towards the front door of the plant. Note the debris inside the entrance.

Fig. 9.79. Ruble, a common scene in the Shifang residential area.

Fig. 9.80. A concrete frame residential building almost tipped over. The structure seems to be in good condition. Most likely the foundation was not properly treated before constructing the building.

9.9 Nanba (南坝) and Hejiaba (何家坝)

The town of Nanba is much like Yingxiu, where most of buildings were damaged. During the team's visit, at least 90 percent of the buildings were damaged, most of them were leveled.

Hejiaba is a very small village across the river from Nanba. The fault that runs under this village ruptured during the earthquake. We noted that buildings close to the fault were all severely damaged.

Fig. 9.81. This area in Nanba, located on the south side of the river, was filled with houses before the earthquake.

Fig. 9.82. The first house we saw on the road to Nanba was located about 2 km from the center of the town. Note that the foundation is just a concrete pad without footing.

Fig. 9.83. This was a common scene in Nanba. Note the concrete building on the right performed quite well.

Fig. 9.84. A collapsed concrete frame building showing the reinforcement bar. The pre-cast floor slabs were resting on beams without structural connection to the concrete frame.

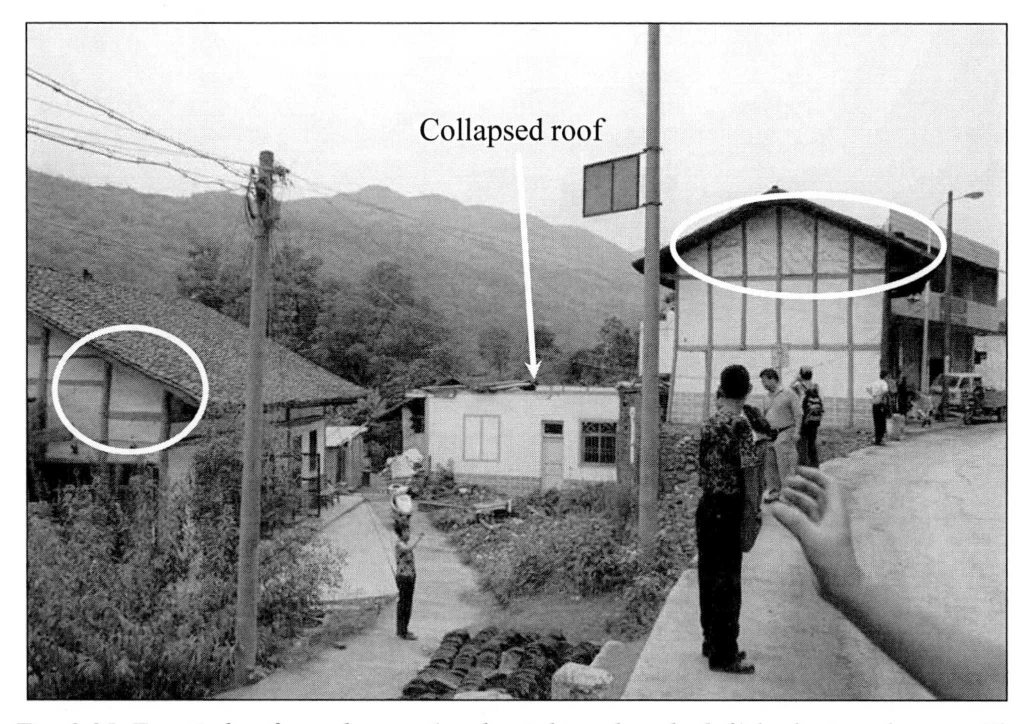

Fig. 9.85. Two timber frame houses (on the right and on the left) had minor damage. The infill exterior walls between the timbers along the roofline failed and were replaced by thin sheet metal to cover the hole. The house in the middle lost its roof.

Fig. 9.86. The traffic police office building on the main road in Nanba collapsed, but the front gate remained standing

Fig. 9.87. The new building on the right, which was still under construction, and the two building facing it did not show any evidence of damage. However, the URM buildings adjacent to them did sustain damage.

Fig. 9.88. In Hejiaba the surface fault crossed this road in the village. The buildings within 200 m of this surface faulting were damaged and many collapsed.

Fig. 9.89. This building is the furthest from the surface rupture, yet it sustained extensive damage and was unsafe to use.

Fig. 9.90. Details of the damage to the ground floor of the building shown in Fig. 9.90

9.10 Lessons Learned and Recommendations

There was much speculation after the earthquake about collapsed and damaged buildings. Accusations of poor workmanship, the use of poor or substandard materials, corruption, flawed design, and overloading the property were commonly reported in the media. From an engineering perspective, two basics must be changed to ensure structural resistance to seismic forces in the future:

1. The seismic zoning of the region must be re-evaluated. The current zoning requirement is too low for the seismicity of the region based on past events.

2. The building code must include seismic design details for both engineered and non-engineered buildings with life safety setting the primary rules of design, with critical facility buildings, hospitals, and schools having a high design load of at least 50 percent more.

Once these basic items are completed, then a process to implement the code and monitor quality (construction and material) and unethical behavior can be deployed. Peer review can be established by local governments to monitor design quality and adherence to standards.

URM buildings can be built to meet the criteria of life safety. There are various design methods available.

Another important lesson is the emergency preparedness of schools. Preparing students to respond to an emergency is critical. Schools must develop and adopt a process to continuously educate students about fire, earthquake, flood, freezing rainstorm, snowstorm, and windstorm emergency response. An annual emergency response simulation should be required for all schools in the region.

9.11 Acknowledgment

The authors are indebted to all the local supports who lead us to the various sites, in particular, Professor Manchao He, Director Zhirong Mei, Director Junwei Cui, and Mr. Jian Wang.

Unless otherwise stated in the caption, all figures and tables belong to the authors of this chapter.

10 EMERGENCY RESPONSE

EXECUTIVE SUMMARY

The Wenchuan Earthquake was a once in a lifetime disaster that affected a large area within the mountainous region of central China. The Chinese government was challenged with the need for immediate response including search and rescue, restoration of lifelines, debris removal, and developing plans for response to this mega event. Special challenges for this event included landslides and rockfalls blocking roads and hindering response times; landslides blocking rivers and diverting response crews to removing blockages to prevent catastrophic earthquake dam failures; rainy weather causing mud slides and debris flows further hindering search and rescue efforts; and post-World War II structures that do not meet current design standards increasing building damage.

Despite these special challenges, the Chinese government responded exceptionally well. It evaluated the conditions, mobilized the army, and coordinated domestic and international relief efforts with speed and prudence. During the course of the ASCE investigations, every person from both Chinese educational and commercial technical institutions was extremely helpful in providing logistical support and transfer of technical information.

10.1 Government Response

Almost immediately after the earthquake, the Chinese government quickly responded to the disaster by establishing a rescue headquarters of the State Council. At the same time leaders of the departments, military units, armed police units, local party committees, and local governments formed eight fighting groups, including the rescue team, monitoring group, medical group, livelihood arrangement group, infrastructure group, production recovery group, security group, and publicity group. Within five hours the emergency response and rescue team was in the Wenchuan area. By midnight, there were more than 20,000 army and police personnel in the disaster area. A team of over 220 medical personnel was sent to Duijiangyan City, where a temporary medical facility was set up in tents.

However, the deployment of the troops was impeded by the fallen bridges, thousands of landslides, and blocked roads (Fig. 10.1). By the second day, many rescuers marched over the landslides into the damaged areas to begin the rescue and relief efforts. On the third day, helicopters landed in Wenchuan, and other helicopters dropped 12.5 tons of food and other relief supplies. Later in the day, 15 soldiers parachuted into Maoxian County and provided the first reports about deaths, injuries, and infrastructure damage. However, the Chinese government did not have enough helicopters, especially the larger ones, to fully deploy into the damage area. A week later, Russia offered its biggest helicopters to deliver heavy construction equipment in the disaster areas to open roads and relief materials.

Fig. 10.1. Large landslide blocking road

Within five days, the Mianyang Food and Drug Administration established a relief team to set up rescue and psychological counseling teams in the affected areas. Pharmaceutical companies were urged to step up production to ensure an adequate supply of medicines and equipment for those in need. The Chinese Ministry of Health also sent US$441,000 worth of medical equipment to 26 disease control centers in the earthquake zones.

Within seven days, the government had deployed more than 100,000 troops, opened two major highways, and sent more than 5,000 medical workers into the damaged areas. Army units and the China Red Cross delivered nearly 300,000 tents (Fig. 10.2 and 10.3), 790,000 blankets, 1,700,000 jackets, and 218 million Yuan worth of food and water. Within the two weeks, the government increased these to more 400,000 tents, 2.3 million blankets, and 3 million jackets. International and domestic donations totaled 21.4 billion Yuan.

Fig. 10.2. Temporary tents

Fig. 10.3. Temporary tents

Rainy weather and the formation of landslide caused lakes created additional challenges for the rescue and restoration efforts. Heavy equipment was flown in the area to construct spillways to open blocked rivers releasing water from behind the dams to prevent catastrophic failures. Heavy rains caused mud flows in the landslide scarred region, which further hindered rescue and relief efforts. Even six months after the event, many roads were still being cleared and many were only open to one way traffic (Fig. 10.4 through 10.9).

Fig. 10.4. Relocated road below blocked road

Fig. 10.5. Relocated road at downed bridge

Fig. 10.6. Road clearing below landslide

Fig. 10.7. One-way traffic and road clearing

Fig. 10.8. Blocked road

Fig. 10.9. Mud/debris flow blocking road

China's civil aviation industry took quick action to ensure smooth traffic in the air after the earthquake. The China Aviation Administration set up a relief leader team and deployed forces to ensure air transportation ran smoothly. They asked commercial airlines to make full efforts to assist and coordinate with the government's rescue and relief efforts. Scheduled flights in and out of Chengdu (Shuang Liu) International Airport were cancelled for about a week to allow relief material and resources reached the victims quickly.

The government also asked the National Tourism Administration to stop organizing tours to the earthquake areas. This allowed the use of hotels and other lodgings for emergency personnel and even catering services for aiding the distribution of food to the needy.

Figures 10.10 and 10.11 show the common styles of temporary housing installed in Yingxiu. The ASCE team observed at least 100 such buildings in Yingxiu, each with 10 living units, suggesting a supported population of a few thousand.

Fig. 10.10. Temporary housing in Yingxiu

Fig. 10.11. Temporary housing in Yingxiu

Even five months after the earthquake, the team observed military groups of 20 to 50 people in each town with high damage rates in the mountainous areas providing support. They were performing guard duty, such as keeping people away from hazardous collapsed building sites and landslide zones.

10.2 Non-Government Organization (NGO) Response

Due to the earthquake's magnitude, the Chinese government was willing to accept assistance from the international community. The United Nations (UN) responded immediately by making a contribution through the Central Emergency Response Fund (CERF). The multi-agency UN Disaster Management Team (UNDMT) coordinated the UN system in China's overall relief and early recovery efforts. As of April 2009, they allocated more than US$8,000,000 to various UNDMT agencies. Throughout the summer, CERF received more than US$10,000,000 from other agencies to augment supplies.

In July 2008, the UN made an appeal for US$33.5 million of aid to assist victims in nine sectors (shelter, health, nutrition and HIV/AIDS, water and sanitation, protection of vulnerable groups, education, livelihoods, environment, ethnic minorities, and coordination). This resulted in donations of more than US$18 million from Canada, Norway, Saudi Arabia, Sweden, Belgium, Finland, the European Union, and Luxemburg.

The International Labor Organization (ILO) provided more than US$1,000,000 in assistance to reestablish more than 1,000 small businesses that were destroyed and set up 700 new ones (Fig. 10.12 to 10.15) for those who lost their jobs.

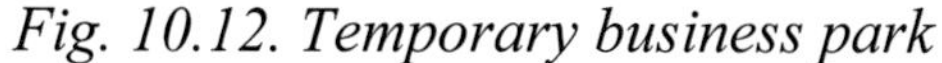

Fig. 10.12. Temporary business park

Fig. 10.13. Temporary offices

Fig. 10.14. Temporary businesses/ homes

Fig. 10.15. Temporary businesses and homes below a May 12 memorial

The UN Educational, Scientific and Cultural Organization (UNESCO) provided training on disaster risk reduction, which included translating educational and emergency related materials and distributing them to the quake-affected areas. In addition, they helped rebuild and restore world heritage sites in Sichuan Province, including the giant panda sanctuaries, the ancient Taoist temple in Mount Qingcheng, and the Duijiangyan irrigation system. They also rehabilitated the television station in Ma'erkang County, Sichuan.

UN-HABITAT delivered 21 prefabricated classrooms (Fig. 10.16) in Gansu Province's Longnan Prefecture's Xihe County to provide classrooms for more than 1,100 children.

Fig. 10.16. Temporary school

In November 2008, the International Federation of Red Cross (IFRC) and Red Crescent Societies provided US$58 million in additional funding to meet survivor's urgent needs and to help fund their relief programs through 2010. The Japanese Red Cross society supported the construction of 4,700 houses in Mianzhu through the IFRC. They also are supporting the reconstruction of 12 elementary and junior high schools as well as the reconstruction of 29 hospitals and 28 clinics in Sichuan, Gansu, and Shaanxi Provinces.

China Mobile's Sichuan branch set up more than 2,100 sites providing free telephone service, free mobile phone battery charging, and free drinking water. This allowed people to keep in touch with their families soon after the event.

10.3 Continuing Efforts by the Chinese Government

In July 2008, the Chinese Ministry of Commerce co-hosted a workshop on post earthquake reconstruction experiences titled "Building Back Better." The workshop developed lessons learned from experiences with other earthquakes and provided strategies for needed assistance to the poor. In August 2008, a draft reconstruction plan was distributed for comment. The final plan was released in November 2008, which outlined a plan for spending in excess of $146.5 billion during the next three years for reconstruction of the damaged areas.

By the end of 2008, the Chinese government distributed more than ¥10.7 billion RMB in aid to nearly 10 million people affected by the earthquake. This included 8.9 million people who lost their homes and/or income source; each received ¥300 RMB per month along with 15 kg of food. The elderly, orphans, and other special cases received ¥600 RMB per month.

To help stabilize the society in the damaged areas, the Chinese government allowed the media to keep the lines of communication open about the disaster. This included prioritizing the installation of thousands of miles of fiber optic lines into the stricken zones.

10.4 Long-Term Reconstruction Efforts

The Chinese government is taking great steps towards the prevention of such monumental losses from potential future disasters. The reconstruction plan includes:

- Reconstruction guidelines noting reconstruction challenges and identifying disaster areas.
- Statements of the basic principles and objectives for the reconstruction effort.
- Discussions on the future land use planning related to urban and rural distribution, industrial distribution, resettlement of population out of hazard areas, and other land use considerations.

- Planning for urban reconstruction, including public utilities and cultural and historic resources.
- Planning for rural reconstruction, including the provision of agricultural services and infrastructure.
- Public-services planning for education, scientific research, health care, culture and sports, cultural and natural heritage, employment and social security, and social management.
- Infrastructure planning for traffic control, communications, energy, and water conservation.
- Specific planning for the reconstruction, including industry, tourism, commerce and trade, finance, and cultural industry.
- Disaster prevention and mitigation planning, including control, mitigation, and relief (emergency response).
- Environmental planning, including restoration, protection, and reclamation.
- Efforts for maintaining the spiritual homeland, including psychological rehabilitation, and maintaining the national spirit.
- Plans for establishing new policies for fiscal reform, taxes, land policies, industry, and assistance.
- Plans outlining fund demands and financing, and fund allocation principles.
- Lastly, planning and implementing the reconstruction efforts, including the leadership and organization, management, implementation, material support and supervision, and inspection.

10.5 Conclusions

Given the magnitude of the earthquake, the widespread damage zone, and the inaccessibility of the damaged areas, the Chinese government's response was better than might be expected (Fig. 10.14 and 10.15). The government's ability to appraise the situation, mobilize relief personnel, and organize the international relief efforts could be used as a model for future disaster responses. Their current efforts to develop a mitigation plan should go along way to prevention of high death tolls in future disaster events.

10.6 Acknowledgment

Much of the data in this chapter was obtained from information posted on the Relief Web at *www.reliefweb.int.*

Index

Page numbers followed by *e*, *f*, and *t* indicate equations, figures, and tables, respectively.